INTRODUCTION TO SYMBOLIC LOGIC

INTRODUCTION TO SYMBOLIC LOGIC

JOHN L. POLLOCK

State University of New York at Buffalo

HOLT, RINEHART and WINSTON, INC.

New York Chicago San Francisco Atlanta
Dallas Montreal Toronto London Sydney

Library of Congress Catalog Card Number: 69-12269

SBN: 03-072765-0

Printed in the United States of America

1 2 3 4 5 6 7 8 9

To Carrot, Klunky, Imbrecilla,
and Question Mark

PREFACE

The propositional and predicate calculi—the backbone of modern logic—would seem to constitute the natural subject matter for a first course in logic, but unfortunately many logic teachers believe them too difficult for beginners. As a result, most logic textbooks either treat the calculi in a completely rigorous manner, without making concessions to the student's inexperience or the instructor's frequent lack of familiarity with the details of the subject (after all, most logic teachers are not logicians), or they attempt to achieve readability and teachability by sacrificing rigor to a degree that is extremely disquieting to the professional logician. In my opinion neither of these courses is necessary. Thus, I have tried to construct a text that, on the one hand, is sufficiently rigorous to satisfy most logicians, but on the other hand, can be read by the uninitiated and taught by the instructor without an extensive background knowledge of logic.

Symbolic Logic begins with a detailed discussion of the propositional calculus, which, although it is not intrinsically interesting, provides an ideal introduction to the basic concepts of logic in a fairly simple setting. This book differs from most texts in logic in that the general approach is semantical. I feel that the student cannot really understand what a deductive system of logic is all about unless the concepts of a valid sentence and of a set of sentences implying another sentence are discussed first. These concepts provide the rationale for the construction of a system of derivations. Although their definitions in the propositional calculus are relatively simple, those in the predicate calculus are much more complex. Since the student is less likely to be confused by the complexity of the definitions if he already has a rough idea of what concepts are being defined, it is best to introduce the concepts in the simpler setting of the propositional calculus.

This book is designed for use in a beginning course in logic, but it can be used equally well in a junior or senior level course in symbolic logic for those whose introduction to logic dealt primarily with traditional logic. The volume contains more material than can be fruitfully discussed in one term, but a basic course can be taught with Chapters I and II and all but section 16 of Chapter III. A slightly more difficult course would include section 16 of Chapter III, which deals with the completeness theorem for the predicate calculus. Either course could be augmented with material from Chapter IV, which consists of a rather cursory introduction to a number of additional topics in logic. The treatment of these topics is intended merely to whet the student's interest and is neither as exhaustive nor as rigorous as the rest of the book. The instructor may wish to lecture on some of these topics himself, but it should also be possible to assign certain sections as required reading without discussing them in class.

I used an earlier version of this book in mimeographed form in my classes for several years, and on the basis of that experience I have one strong recommendation for the instructor: during examinations the students should be allowed to use lists of the equivalences and implications necessary for constructing derivations. It seems useless to require students to memorize the code numbers for 17 implication schemes and 27 equivalence schemes, particularly since those code numbers are idiosyncratic to this book. Allowing the students to use such a list will not seriously compromise the extent to which they must be familiar with its contents. During a timed examination, if the student does not know what scheme he is looking for, a list of the 44 schemes will not help him.

It may be felt that the use of 44 implication and equivalence schemes in the construction of derivations is excessive. Of course, the list can be cut drastically—the extent to which it can be cut is discussed in the exercises. However, I have included all the schemes in order to make derivations seem as natural as possible and to make them reflect the structure of our informal reasoning as closely as possible.

The book contains two categories of exercises. Most sections are followed by exercises whose satisfactory solution should be within the powers of the average student. In addition, some sections are followed by more difficult exercises, indicated by asterisks. Many of these can only be solved by the exceptional student. A few—those that do not have generally accepted solutions—reflect philosophical problems that arise in the foundations of logic.

This text owes a great debt to a number of people without whom it would never have been written. Foremost among these are Ernest W. Adams, from whom I learned to teach logic, and Benson Mates, whose book *Elementary Logic* was an inspiration to me. I also owe much to my students, who suffered through numerous experiments concerned with how best to teach the predicate calculus to freshmen and sophomores, and finally to my wife, who read the manuscript from the point of view of a beginner in logic.

J. L. P.

Buffalo, New York
May 1969

CONTENTS

III. THE PREDICATE CALCULUS

IV. OTHER TOPICS IN LOGIC

INTRODUCTION TO SYMBOLIC LOGIC

I INTRODUCTION

I.1 LOGIC AS THE STUDY OF SOUND ARGUMENT FORMS

Logic can be thought of as the study of arguments. In logic we try to say what makes an argument a good argument.

First then, what is an argument? An argument begins with a finite set of statements that are the *premises* of that argument. Then it proceeds to draw conclusions from these premises until it arrives at the last statement, which is the *conclusion* of the argument. The following is an example of an argument:

PREMISES

A. 1. If we continue to send men and arms to Vietnam, our stock piles of raw materials will be seriously depleted.
2. If our stock piles of raw materials are seriously depleted, we won't be able to defend the rest of Southeast Asia from Communist aggression.
3. If we can't defend the rest of Southeast Asia from Communist aggression, Communist China will take over much of Southeast Asia.
4. But if we stop sending men and arms to Vietnam, the Viet Cong will take over the country.
5. If the Viet Cong take over Vietnam, Communist China will be able to use Vietnam as a springboard for conquering much of the rest of Southeast Asia.

INTERMEDIATE INFERENCES

6. Thus if we continue to send men and arms to Vietnam, we won't be able to defend the rest of Southeast Asia from Communist aggression. (From 1 and 2)

7. Thus if we continue to send men and arms to Vietnam, Communist China will take over much of Southeast Asia. (From 6 and 3)
8. But if we stop sending men and arms to Vietnam, Communist China will be able to use Vietnam as a springboard for conquering much of the rest of Southeast Asia. (From 4 and 5)

CONCLUSION

9. Thus no matter what we do, Communist China will conquer much of Southeast Asia.

Lines 1–5 are the premises of this argument, lines 6–8 the *intermediate inferences* from the premises, and line 9 the conclusion. Intermediate inferences are the steps of an argument between the premises and the conclusion. Often we have to use intermediate inferences to help us draw our final conclusion. However, arguments like the following two are very simple ones that do not require intermediate inferences:

B.
1. The moon is made of green cheese.
2. If the moon is made of green cheese, then $2 + 2 = 4$.

---

3. Therefore, $2 + 2 = 4$.

C.
1. All featherless bipeds are men.
2. All men are mortal.

---

3. Therefore, all featherless bipeds are mortal.

An argument is said to be *sound* (or *valid*) when the conclusion follows from, or is a consequence of, the premises. The above three arguments are all sound because if their premises are true, their conclusions must also be true. It should be emphasized that it is not necessary for the premises of an argument to be true in order for the argument to be sound. For example, Argument B has a false premise, and many of the premises in Argument A are at least questionable, but these are still sound arguments. All that is necessary for an argument to be sound is that there be some relationship between the premises and the conclusion so that if the premises *were* all true then the conclusion *would be* true. Another way to put this is to say that in a sound argument it is impossible for all the premises to be true and the conclusion to be false.

It should also be emphasized that if the premises of a sound argument are false, it does not follow that the conclusion is also false. The following is an example of a sound argument with (presumably) false premises and a true conclusion:

D. 1. All Communists are members of the Republican Party.
2. All members of the Republican Party are followers of Karl Marx.

---

3. Therefore, all Communists are followers of Karl Marx.

Given a sound argument, the only time we can infer anything about whether its conclusion is true is when all its premises are true. If one or more of its premises are false, then its conclusion can be *either* true or false. The *only* combination of truth and falsity that is prohibited in a sound argument is all true premises and a false conclusion. Any other combination is possible.

In Arguments B, C, and D we can see immediately that their conclusions follow from their premises just by looking at them. No intermediate inferences are required. Perhaps this is also true of Argument A, although it is a bit more complicated. But there will be many cases in which, given a set of premises and a conclusion, one cannot tell just by looking at them whether the conclusion follows from the premises. For example, consider the following set of three premises and a conclusion:

E. 1. If the river floods, then our entire wheat crop will be destroyed.
2. The river will flood if, and only if, there is an early thaw in the mountains.
3. But in any case there will be heavy rains later in the summer.

---

4. Therefore, if there is an early thaw in the mountains, then our entire wheat crop will be destroyed, and the community will be bankrupt unless there are heavy rains later in the summer.

This is a sound argument, but it is not one that seems to be sound just by looking at it. Unlike our earlier examples, this argument cannot be evaluated intuitively. What logic does is provide us with methods for evaluating the soundness of more complex arguments like this. Insofar as it does this, it is clearly a useful tool for any intellectual endeavor. Of course, given the above argument and an unlimited amount of time, a fairly intelligent person who was completely untutored in logic could probably convince himself that the argument is sound. Similarly, a bank teller who knew no arithmetic might be able to make change by counting on his toes, but how much simpler it is to be able to do sums in your head. And analogously, college students faced with fairly complex arguments will be much better off knowing some of the basic laws of logic, which immediately make clear that the above argument is sound.

In logic we are interested in seeing what general features of an argument make it sound. We can draw a useful parallel with physics. The physicist is not much interested in the motion of any particular physical object. Rather, he is interested in those laws that describe the motion of all physical objects. Similarly, in logic we are not much interested in particular arguments. Rather, we are interested in those general features that many arguments may have in common. This leads us to talk about the *forms* of arguments. For example, the form of Argument D is:

1. All *A* are *B*.
2. All *B* are *C*.

---

3. Therefore, all *A* are *C*.

This argument form is interesting because *any* argument that has this same form will be sound. For example, the argument

F. 1. All dogs are warm-blooded animals.
2. All warm-blooded animals have central nervous systems.

---

3. Therefore, all dogs have central nervous systems.

is also of this form, and it is also sound. The fact that an argument has this form is sufficient to *make* it a sound argument. In logic our principal interest is to investigate those argument forms that have this latter characteristic; that is, we are interested in those argument forms that are such that any argument of that form will be sound. We will say that the argument form itself is sound if every individual argument having that form is sound.

In talking about argument forms we are led to "laws of logic." For example, it is a law of logic that any argument having the form

1. All *A* are *B*.
2. All *B* are *C*.

---

3. Therefore, all *A* are *C*.

is sound. Laws of logic allow us to express general truths about arguments just as laws of physics allow us to express general truths about physical objects.

1.2 LOGIC AS THE STUDY OF FORMAL ANALYTICITY

Logic can also be thought of in another way. Consider the statement "All bachelors are unmarried." This can be seen to be true simply by reflecting upon what it means because part of what it means to say that someone is a bachelor is that he is unmarried. Such a statement is said to be *analytic*. An analytic statement is a statement that is true by virtue of the meanings of the terms used in it. Another example of an analytic statement is "All brothers are male," because it is part of the meaning of "brother" that a brother is male. An analytic statement is one that could not possibly be false. An analytic statement is *necessarily true*—true in all possible worlds.

To be contrasted with analytic statements are such statements as "There are four apples in this box." The only way to find out whether the latter statement is true is to look in the box and count the apples. You cannot determine whether it is true simply by considering what it means. A statement that is not analytic is *synthetic*.[1]

In logic we are interested in analytic statements. But once again, we are not interested in particular analytic statements as such. Rather, we are interested in those general features of statements that may make them analytic. Thus, just as we talked about the forms of arguments, we are led to talk about the *forms* of statements. Sometimes a statement will have a form such that *any* statement having that same form will be analytic. Such a statement is said to be *formally analytic*. Consider for example the statement "Either it is raining or it is not the case that it is raining." This statement is analytic because you can tell that it is true just by considering what it means. You do not have to know anything about the weather. Similarly, the statements "Either it is snowing or it is not the case that it is snowing," "Either Socrates was bald or it is not the case that Socrates was bald," and "Either $2 + 2 = 4$ or it is not the case that $2 + 2 = 4$" are all analytic. These statements are all analytic for the same reason. They all have the form "Either P or it is not the case that P," where we put some particular statement in for P. Any statement of this form will be

[1] A further discussion of analyticity may be found on pages 230–233.

analytic, because it is of this form. Thus all these statements are formally analytic.

Not all analytic statements are formally analytic. The statement "All bachelors are male" is analytic, but it is not formally analytic. Its form is "All A are B," but there are many statements of this form, such as "All men have livers" or "All bachelors are female," which are not analytic. Thus it is not the *form* of the statement "All bachelors are male" that makes it analytic, and so it is not formally analytic.

We have seen what it means to say that a statement is formally analytic. Now let us talk about statement *forms*. We say that a statement form is formally analytic if and only if it is the form of a formally analytic statement. "Either P or it is not the case that P" is a formally analytic statement form.

Formally analytic statement forms can be thought of as expressing laws of logic. They tell us about the "logical structure of the universe." For example, the statement form "Either P or it is not the case that P" is formally analytic. It is called the Law of the Excluded Middle (there is no middle ground between truth and falsity). This law can be thought of as putting constraints on the way the universe must be. For example, it tells us that regardless of what the weather is, it must either be raining or not be raining. Logic can then be thought of as the study of these formally analytic statement forms.

I.3 THE TWO CONCEPTIONS OF LOGIC

We now have two conceptions of logic—logic as the study of sound argument forms, and logic as the study of formal analyticity. The most important connection between them is the following. Consider an argument having some number, say n, premises: $P_1, P_2, P_3, \ldots, P_n$. Let us suppose the conclusion is some statement, Q. Then the argument looks as follows:

$$
\begin{array}{l}
P_1 \\
P_2 \\
P_3 \\
\vdots \\
P_n \\
\hline
\text{Therefore, } Q.
\end{array}
$$

Corresponding to this argument we can construct the statement: If P_1 and P_2 and P_3 and . . . P_n, then Q. This statement says that if the premises of the above argument are all true then the conclusion is true. This statement is called the *corresponding conditional* of the argument. A *conditional* is any statement of the form "If P then Q." The corresponding conditional of an argument is the conditional in which P says that the premises of the argument are all true, and Q is the conclusion of the argument. For example, the corresponding conditional of Argument C is

> If all featherless bipeds are men, and all men are mortal, then all featherless bipeds are mortal.

To say that an argument is sound is to say that it is necessarily true that if its premises are all true then its conclusion is true; that is, its corresponding conditional is necessarily true, or analytic. Thus, for example, to say that Argument C is sound is just to say that the statement "If all featherless bipeds are men, and all men are mortal, then all featherless bipeds are mortal" is analytic. Thus we can talk about the soundness of an argument by talking about the analyticity of its corresponding conditional: *An argument is sound if, and only if, its corresponding conditional is analytic.*

Similarly, we can talk about the soundness of an argument *form* by talking about the formal analyticity of statement *forms*. Consider an argument form

$$\begin{array}{l} P_1 \\ P_2 \\ P_3 \\ \vdots \\ P_n \\ \hline \text{Therefore, } Q. \end{array}$$

where $P_1, P_2, P_3, \ldots, P_n$, and Q are now statement forms rather than statements. Recall that an argument *form* is sound when, and only when, every particular argument of that form is sound. But because an argument is sound just when its corresponding conditional is analytic, the above argument *form* is sound if, and only if, every conditional having the form "If P_1 and P_2 and P_3 and . . . and P_n, then Q" is analytic. That is, the argument form is sound if, and only if, this corresponding conditional is formally analytic.

Thus formal analyticity is another way to talk about sound argument forms. *An argument form is sound if, and only if, its corresponding conditional is formally analytic.* Thus, the argument form

All *A* are *B*.
All *B* are *C*.

Therefore, all *A* are *C*.

is sound, because the corresponding conditional "If all *A* are *B* and all *B* are *C*, then all *A* are *C*" is formally analytic—regardless of what we substitute for *A*, *B*, and *C*, we get an analytic statement. This, then, is the most important connection between the two conceptions of logic. It gives us a way of talking about sound argument forms in terms of the formal analyticity of statements.

II THE PROPOSITIONAL CALCULUS

II.1 THE SENTENTIAL CONNECTIVES

There is a large class of statements and arguments whose forms can be expressed adequately using the expressions "and," "or," "if . . . then," "if and only if," and "it is not the case that." These expressions are called the *sentential connectives*, because they are used to connect sentences to form larger sentences. For example, the statement "Either it is raining or it is not the case that it is raining" has the form "P or it is not the case that P," where now P stands for "It is raining." Similarly, the statement "If John goes to the party, then Joe will stay home" has the form "If P then Q," where P stands for "John will go to the party" and Q stands for "Joe will stay home."

Unfortunately, in natural languages, such as English, statements that have the same form do not always look the same. For example, the statements "Either it is snowing or it is not the case that it is snowing" and "Either it is sleeting or it is not sleeting" have the same form; they are both formed by saying that either one statement is true or else a second statement is true, where the second statement simply denies the truth of the first statement. But they do not look quite the same because the "not" is placed differently in the two statements. It is therefore advantageous to introduce an artificial symbolism to represent the form of a statement. We replace "not" by "$\sim$," and then both "It is not the case that it is sleeting" and "It is not sleeting" can be represented by "$\sim$ (it is sleeting)."

Letting P and Q be arbitrary statements, we can introduce the following symbols:

Symbol	*Meaning*
$\sim P$	It is not the case that P.
$(P \,\&\, Q)$	P and Q (or, both P and Q).
$(P \vee Q)$	P or Q (or, either P or Q).
$(P \supset Q)$	If P then Q.
$(P \equiv Q)$	P if and only if Q.

These symbols are abbreviations for the sentential connectives.[1] Statements formed using these symbols have names, as follows: $\sim P$ is the *negation* of P; $(P \,\&\, Q)$ is the *conjunction* of P and Q, and P and Q are the *conjuncts*; $(P \vee Q)$ is the *disjunction* of P and Q, and P and Q are the *disjuncts*; $(P \supset Q)$ is a *conditional*, and P is the *antecedent*, and Q is the *consequent*; $(P \equiv Q)$ is a *biconditional*.

Using these symbols we can represent the forms of quite complex statements. We have been using capital letters to symbolize the simple statements that form the parts of compound statements. We might occasionally need more than 26 letters, so we can also use capital letters with subscripts, such as P_7 or Q_{13}. It should be emphasized that the same letter with different subscripts, such as P and P_7, can be used to represent totally unrelated statements. The fact that P occurs in P_7 is just a coincidence. Capital letters, with or without subscripts, used this way are called *sentential letters*. In the statement "Either it is raining or it is not the case that it is raining," we can let P_3 be the statement "It is raining." Then the above statement becomes "Either P_3 or it is not the case that P_3." Using the symbols for the sentential connectives, this in turn can be symbolized as "Either P_3 or $\sim P_3$," and then as $(P_3 \vee \sim P_3)$.

Now consider some more examples of symbolizing the forms of statements like "If he comes we will have the party at his house, and if he doesn't come then we will have the party at Jones' house." Letting P be the statement "He comes," Q the statement "We will have the party at his house," and R the statement "We will have the party at Jones' house," we can first symbolize the statement as "If P then Q, and if it is not the case that P then R." This in turn can be symbolized as "If P then Q, and if $\sim P$ then R," and then as "$(P \supset Q)$ and $(\sim P \supset R)$," and finally as $[(P \supset Q) \,\&\, (\sim P \supset R)]$.

[1] In what follows, these symbols will be called the sentential connectives, although strictly speaking they are abbreviations for the sentential connectives. The sentential connectives themselves are English words and phrases.

As another example, consider "If Jones needs money, then either he will reduce prices or he will apply to the bank for a loan." Letting P be "Jones needs money," Q be "He will reduce prices," and R be "He will apply to the bank for a loan," this statement can be symbolized, first as "If P then either Q or R," then as "If P then $(Q \vee R)$," and finally as $[P \supset (Q \vee R)]$.

There are two things that should be noticed about the preceding examples. The first is that in symbolizing relatively complex statements like "If he comes we will have the party at his house, and if he doesn't come then we will have the party at Jones' house," we begin by symbolizing the smallest parts, and then we construct the successively larger parts one at a time until we finally get the whole statement. The symbolization went as follows:

$$\underbrace{\text{If } P \text{ then } Q,}\ \text{and}\ \underbrace{\text{if it is not the case that } P \text{ then } R}$$

$$(P \supset Q)\ \text{and}\ \underbrace{\text{if}\quad \sim P \quad \text{then } R}$$

$$\underbrace{(P \supset Q)\ \text{and}\quad (\sim P \supset R)}$$

$$[(P \supset Q)\ \&\ (\sim P \supset R)]$$

It is always best to begin by symbolizing the smallest parts of a statement first, and then constructing the symbolization of the successively larger parts one at a time in terms of the parts that make them up. It is unwise to try to symbolize an entire statement in one fell swoop if the statement is at all complicated.

Second, notice the use of parentheses. Each time we symbolize a part of the statement—unless that part is a negation—we enclose it in parentheses to keep it separate from the other parts of the statement. The parentheses take the place of commas and other grammatical conventions of English. We must always be careful to put the parentheses in, or the resulting symbolization will be ambiguous. Consider the two statements "It is not true that Jones has a girlfriend and his wife is going to divorce him" and "It is not true that Jones has a girlfriend, and his wife is going to divorce him." These obviously mean quite different things. The first denies that it is true both that Jones has a girlfriend and that his wife is going to divorce him, whereas the second says that Jones does not have a girlfriend, but his wife is going to divorce him anyway. These two statements would be symbolized as $\sim(P \,\&\, Q)$ and $(\sim P \,\&\, Q)$ respectively. But if we omitted the parentheses in symbolizing these statements and just wrote $\sim P \,\&\, Q$, we would not know which of these two different statements were meant.

It is not necessary to enclose a negation (a statement of the form $\sim P$) in parentheses, because "$\sim$" does not really connect sentences—it acts on a single sentence. But for any of the other sentential connectives it is necessary to use parentheses each time the connective is used.

II.1 EXERCISES

Symbolize the forms of the following statements:

1. If it rains today the ground will be wet, and we will not be able to have a picnic.
2. It is not true that if it is cloudy then it will rain.
3. If the sun shines in the morning it will rain, and if the sun does not shine in the morning then it will not rain.
4. If we get plenty of sunshine, then if it rains the flowers will grow.
5. It is not the case that, there is a woman in the next room if and only if Jim said there is.
6. It is not the case that there is a woman in the next room, if and only if, Jim said there is.
7. Either the entrails will contain cockroaches or they will not, and if they do then the gods are angered, and if they do not then Venus will be in apposition to Jupiter.
8. If she is an acrobat or she is a clown, then she lives in that trailer.
9. Harry will go to the bank today if, and only if, the market drops, and if Harry goes to the bank today the supervisors will hide, and if the supervisors hide and Harry does not go to the bank today, then Emmett will lose his shirt on the stock market.
10. If Francis Bacon wrote *Hamlet* and Shakespeare wrote *Macbeth*, then either Shakespeare was Bacon, or the theater manager was a crook.

II.2 FORMULAS OF THE PROPOSITIONAL CALCULUS

Using sentential letters to stand for statements, and then combining them with sentential connectives, we can construct quite complex statement forms, called *formulas of the propositional calculus*. We can give precise rules for constructing formulas of the propositional calculus. The simplest formulas are simply sentential letters, with or without subscripts, such as P, Q, P_{13}, R_{129}, and so forth. Let us call these *atomic formulas of the propositional calculus* because they are the atoms from which more complicated formulas are constructed. Then all other formulas of the propositional calculus can be constructed by successive applications of the following rules:

1. An atomic formula is a formula.
2. If P is any formula, then $\sim P$ is a formula.
3. If P and Q are any formulas, then $(P \& Q)$ is a formula.
4. If P and Q are any formulas, then $(P \vee Q)$ is a formula.
5. If P and Q are any formulas, then $(P \supset Q)$ is a formula.
6. If P and Q are any formulas, then $(P \equiv Q)$ is a formula.

All formulas of the propositional calculus can be constructed by repeated application of these rules. Consider the formula

$$(P \supset (\sim Q \& R))$$

We can build this formula up using the above six rules as follows. We begin with the smallest parts and work outwards. By Rule 1, P, Q, and R are formulas. Then by Rule 2, $\sim Q$ is a formula. By Rule 3, as $\sim Q$ and R are both formulas, $(\sim Q \& R)$ is a formula. Then by Rule 5, as P and $(\sim Q \& R)$ are both formulas, $(P \supset (\sim Q \& R))$ is a formula.

We can construct very complicated formulas using these rules. For example, consider the formula

$$((P \supset (Q_3 \equiv \sim R_4)) \\ \equiv \sim\sim(\sim P \& \sim(Q_3 \equiv (R_4 \vee (R_{17} \& \sim\sim P)))))$$

Let us see how we would build this formula up using the six rules. Again, we begin with the smallest parts and work outwards. By Rule 1, P, Q_3, R_4, and R_{17} are formulas, because they are atomic formulas. As P and R_4 are formulas, we can use Rule 2 to construct the formulas $\sim P$ and $\sim R_4$. As $\sim P$ is then a formula, by Rule 2 again, $\sim\sim P$ is a formula. Then as R_{17} and $\sim\sim P$ are both formulas, by Rule 3, $(R_{17} \& \sim\sim P)$ is a formula. Then as R_4 and $(R_{17} \& \sim\sim P)$ are both formulas, by Rule 4, $(R_4 \vee (R_{17} \& \sim\sim P))$ is a formula. Then as Q_3 and $(R_4 \vee (R_{17} \& \sim\sim P))$ are both formulas, by Rule 6,

$$(Q_3 \equiv (R_4 \vee (R_{17} \& \sim\sim P)))$$

is a formula. Then by Rule 2,

$$\sim(Q_3 \equiv (R_4 \vee (R_{17} \& \sim\sim P)))$$

is a formula. We have already seen that $\sim P$ is a formula, so by Rule 3,

$$(\sim P \;\&\; \sim(Q_3 \equiv (R_4 \vee (R_{17} \;\&\; \sim\sim P))))$$

is a formula. Then by Rule 2,

$$\sim(\sim P \;\&\; \sim(Q_3 \equiv (R_4 \vee (R_{17} \;\&\; \sim\sim P))))$$

is a formula, and then by Rule 2 again,

$$\sim\sim(\sim P \;\&\; \sim(Q_3 \equiv (R_4 \vee (R_{17} \;\&\; \sim\sim P))))$$

is a formula. This gives us the right side of the biconditional. Now working on the other side, as Q_3 and $\sim R_4$ are both formulas, by Rule 6, $(Q_3 \equiv \sim R_4)$ is a formula. P is a formula, so by Rule 5, $(P \supset (Q_3 \equiv \sim R_4))$ is a formula. Both sides of the biconditional have now been constructed, and then by Rule 6,

$$((P \supset (Q_3 \equiv \sim R_4)) \\ \equiv \sim\sim(\sim P \;\&\; \sim(Q_3 \equiv (R_4 \vee (R_{17} \;\&\; \sim\sim P)))))$$

is a formula. We can diagram the construction of this formula as follows:

$P \quad Q_3 \quad R_4 \quad P \quad Q_3 \quad R_4 \quad R_{17} \quad P$	Rule 1
$\sim R_4 \quad \sim P \quad \sim P$	Rule 2
$\sim\sim P$	Rule 2
$(R_{17} \;\&\; \sim\sim P)$	Rule 3
$(R_4 \vee (R_{17} \;\&\; \sim\sim P))$	Rule 4
$(Q_3 \equiv (R_4 \vee (R_{17} \;\&\; \sim\sim P)))$	Rule 6
$\sim(Q_3 \equiv (R_4 \vee (R_{17} \;\&\; \sim\sim P)))$	Rule 2
$(\sim P \;\&\; \sim(Q_3 \equiv (R_4 \vee (R_{17} \;\&\; \sim\sim P))))$	Rule 3
$\sim(\sim P \;\&\; \sim(Q_3 \equiv (R_4 \vee (R_{17} \;\&\; \sim\sim P))))$	Rule 2
$\sim\sim(\sim P \;\&\; \sim(Q_3 \equiv (R_4 \vee (R_{17} \;\&\; \sim\sim P))))$	Rule 2
$(Q_3 \equiv \sim R_4)$	Rule 6
$(P \supset (Q_3 \equiv \sim R_4))$	Rule 5
$((P \supset (Q_3 \equiv \sim R_4)) \equiv \sim\sim(\sim P \& \sim(Q_3 \equiv (R_4 \vee (R_{17} \& \sim\sim P)))))$	Rule 6

Notice that whenever we construct a conjunction, disjunction, conditional, or biconditional, we enclose it in parentheses. In order to make formulas easier to read, parentheses are often replaced by brackets. Thus we might write the above formula as

$$\{[P \supset (Q_3 \equiv \sim R_4)] \\ \equiv \sim\sim[\sim P \;\&\; \sim(Q_3 \equiv \{R_4 \vee (R_{17} \;\&\; \sim\sim P)\})]\}$$

By doing this we can tell at a glance which left parenthesis or bracket goes with which right parenthesis or bracket; thus, we can see more easily what the formula means.

Atomic formulas are the simplest formulas. Let us say that a formula is *molecular* if it is not atomic. Then molecular formulas are those formulas that contain sentential connectives.

Sometimes we will want to talk about one formula being a *part of* another formula. This just means that the first occurs somewhere in the second. For example, P, Q, R, $(P \supset Q)$, $\sim(P \supset Q)$, and

$$[\sim(P \supset Q) \vee R]$$

are all parts of the formula $\sim[\sim(P \supset Q) \vee R]$. Let us also adopt the convention of saying that a formula is part of itself. Then

$$\sim[\sim(P \supset Q) \vee R]$$

is also a part of $\sim[\sim(P \supset Q) \vee R]$. The *atomic parts* of a formula are those parts of the formula that are atomic. Thus P, Q, and R are the atomic parts of $\sim[\sim(P \supset Q) \vee R]$, while $(P \supset Q)$, $\sim(P \supset Q)$, $[\sim(P \supset Q) \vee R]$ and $\sim[\sim(P \supset Q) \vee R]$ are parts that are not atomic.

II.2 EXERCISES

A. Show how the following formulas of the propositional calculus can be constructed using Rules 1 through 6. Diagram the construction as was done on page 14.

1. $((P \supset Q) \equiv (\sim(Q \supset P) \vee \sim\sim(Q \,\&\, \sim P)))$
2. $\sim((Q \,\&\, R) \supset \sim(\sim Q \,\&\, (R \supset (Q \vee \sim R))))$
3. $((P \vee (Q \,\&\, R)) \equiv \sim(\sim P \,\&\, (\sim Q \vee \sim R)))$

B. Construct a table for each of the following expressions indicating: (1) whether it is a formula of the propositional calculus; (2) if it is a formula, whether it is atomic or molecular; (3) if it is a molecular formula, whether it is a negation, conjunction, disjunction, conditional or biconditional; (4) if it is a formula, what its atomic parts are; (5) if it is a formula, what its parts are:

1. P
2. $\sim\sim P$
3. $P \vee Q$
4. $((P \vee Q) \supset \sim Q)$
5. $(\sim(P \equiv Q))$
6. $(P \vee Q \supset R)$
7. $\sim((Q \,\&\, R) \equiv (\sim Q \vee \sim R))$
8. $(P \vee P)$
9. $(P \,\&\, \sim P)$
10. $(P \supset Q) \equiv \sim(Q \supset P)$

II.3 PARAPHRASING

Sometimes it is necessary to paraphrase a statement before it can be symbolized. Consider the statement "John and Joe both came to the party." We want to symbolize this as a conjunction, but it cannot be symbolized directly because "and" stands between two names rather than between two sentences. Before we can symbolize it we must paraphrase it so that the sentential connective connects two sentences, "John came to the party and Joe came to the party." Then letting P be "John came to the party" and Q be "Joe came to the party," we can symbolize it as $(P \mathbin{\&} Q)$.

Another example of such paraphrasing is found in the following statement: "If either Kennedy or Khrushchev had been weaker willed concerning either Berlin or Cuba, the cold war would have turned into a hot war." This must be paraphrased first as "If either Kennedy had been weaker willed concerning either Berlin or Cuba, or Khrushchev had been weaker willed concerning either Berlin or Cuba, then the cold war would have turned into a hot war." Then this must be paraphrased again as "If either Kennedy had been weaker willed concerning Berlin or Kennedy had been weaker willed concerning Cuba, or Khrushchev had been weaker willed concerning Berlin or Khrushchev had been weaker willed concerning Cuba, then the cold war would have turned into a hot war." Finally then, this can be symbolized as $\{[(P \vee Q) \vee (R \vee S)] \supset T\}$.

Other kinds of paraphrasing may also be necessary. There are expressions in English such as "unless," "but," "if," "only if," "neither . . . nor," that are much like the sentential connectives. Whenever these occur in a statement, the statement must be paraphrased to replace them by sentential connectives. For example, "Neither Joe came to the party nor John came to the party" means the same thing as "Joe didn't come to the party and John didn't come to the party," and so it must be paraphrased in that way. In general, "Neither P nor Q" can be paraphrased as $({\sim}P \mathbin{\&} {\sim}Q)$.

"But" is like a forceful "and." We may say "He came but he didn't like it" rather than "He came and he didn't like it" merely to emphasize the second conjunct. This emphasis makes no difference for logic, so we can symbolize "P but Q" as simply $(P \mathbin{\&} Q)$.

The behavior of the expressions "if" and "only if" is somewhat surprising. There seems to be a strong temptation to identify "P if Q"

with "If P then Q," but in fact it should be the other way around; that is, "P if Q" means the same thing as "If Q then P." Consider the statement "The crops will be destroyed if there is a flood." To say this is not to say that the crops might not be destroyed anyway, for example, by a drought. But the statement "If the crops are destroyed there is a flood" precludes the possibility of them being destroyed by a drought, therefore it cannot be a proper paraphrase of "The crops will be destroyed if there is a flood." The proper paraphrase is "If there is a flood then the crops will be destroyed." In general, "P if Q" can be symbolized as $(Q \supset P)$.

"P only if Q" works just the other way around. It means "If P then Q." Consider the statement "The crops will be destroyed only if there is a flood." This means that the only way the crops can be destroyed is by a flood, and hence if the crops *are* destroyed then there must have been a flood. This then means "If the crops are destroyed then there is a flood."

Note that "P if and only if Q" is just the conjunction of "P if Q" and "P only if Q," and thus that $(P \equiv Q)$ is equivalent to

$$[(P \supset Q) \,\&\, (Q \supset P)].$$

One further expression that can be paraphrased in terms of the sentential connectives is "unless." "P unless Q" can be paraphrased as $(\sim Q \supset P)$. Suppose we want to paraphrase the statement "We will go to the beach unless it rains." This is the same thing as saying "If it doesn't rain then we will go to the beach;" that is $(\sim Q \supset P)$.

II.3 EXERCISES

Symbolize the forms of the following statements:

1. Neither Jack nor Jim will come unless Mary comes.
2. We will not get there on time unless we speed, but if we speed we might not get there at all.
3. We can get the door open only if we use an acetylene torch on it, but then the door will be ruined.
4. The river will not overflow its banks unless we either have an early thaw or heavy rains, but we will not have heavy rains.
5. Unless we have a flat tire, we can get there on time if we speed, but we will have a flat tire if we speed.

6. Neither Jack nor Jim will come if Mary comes, unless Joan and Mary both come.
7. Jeremy will get a Mercedes Benz for Christmas only if he does not offend Santa Claus, but Jeremy will offend Santa Claus if he does not believe in him, and Jeremy does not believe in Santa Claus.
8. Rain is imminent.
9. John will not come unless Jim comes, and Jim will not come if Jeffrey comes, but Jeffrey will only come if John does not come.
10. It will rain if the barometer drops, but if it rains it will cool off later, and it will not cool off later.

II.4 DIVERGENT USES

One thing to beware of in symbolizing the forms of statements is that there are divergent uses of some of the sentential connectives in which they do not have their ordinary meaning. For example, "and" sometimes means "and then" rather than simply "and." In its ordinary use, "*P* and *Q*" means "It is true that *P*, and it is true that *Q*." On this reading, "*P* and *Q*" means the same thing as "*Q* and *P*." For example, "This is red and that is white" means the same thing as "That is white and this is red." But consider the use of "and" in the following sentence: "She got married and had a baby." This clearly does not mean the same thing as "She had a baby and got married." The former means "She got married and then had a baby." This use of "and" cannot be symbolized simply as "&." The same thing is true of "and" in "He lay down and fell asleep."

The sentential connective that has the greatest number of divergent uses is "if... then." What will be called the standard use of "if... then" is that in which "If *P* then *Q*" is true under the same conditions as "Either $\sim P$ or *Q*"; that is, $(\sim P \vee Q)$. It is only this meaning of "If *P* then *Q*" that can be symbolized as $(P \supset Q)$. First it should be seen that "if... then" is at least sometimes used in this way. Consider the statement "If it rains then the crops will grow." Let us symbolize this as "If *P* then *Q*," where *P* means "It will rain" and *Q* means "The crops will grow." Now it will be argued that this is true under exactly the same circumstances as "Either it won't rain, or the crops will grow"; that is, $(\sim P \vee Q)$:

1. Suppose that "If *P* then *Q*" is true. So if it rains (if *P* is true) then *Q* is true. If it does not rain, then $\sim P$ is true. But either it will

not rain or it will rain. So either it will not rain, and then $\sim P$ will be true, or else it will rain, and then Q will be true. Thus, no matter what happens, either $\sim P$ will be true or Q will be true; that is, $(\sim P \vee Q)$ will be true. So we see that if the statement "If P then Q" is true, then $(\sim P \vee Q)$ is true.

2. In order to show that "If P then Q" and $(\sim P \vee Q)$ are true under exactly the same circumstances, the converse of the above must now be proven, namely, that if $(\sim P \vee Q)$ is true then "If P then Q" is true. So let us suppose that $(\sim P \vee Q)$ is true. A disjunction "A or B" is true if, and only if, at least one of its disjuncts is true; that is, if, and only if, either A is true or B is true. Thus if $(\sim P \vee Q)$ is true, then either $\sim P$ is true or Q is true. Now it will be shown that "If P then Q" is true. If P is true, then $\sim P$ cannot be true. But either $\sim P$ or Q must be true. So if P is true then Q must be true; that is, "If P then Q" is true. Thus it has been shown that if $(\sim P \vee Q)$ is true, then "If P then Q" is true.

Hence the two statements, "If it rains the crops will grow," and "Either it won't rain or the crops will grow," are true in exactly the same circumstances. In this context "if... then" has what is called its "standard use."

However, there are other uses of "if... then" in which "if P then Q" is clearly not the same as "Either $\sim P$ or Q." One place this occurs is in what are called *counterfactual conditionals*. These are statements such as "If this match had been struck it would have lit," which tell us that if something that did not happen *had* happened, then something else would have been the case. To see that these conditionals cannot be translated into disjunctions, let us suppose that we have a match which was not struck. Then exactly one of the following two statements is true: "If this match had been struck it would have lit"; "If this match had been struck it wouldn't have lit." If "if... then" had its standard use in these statements, they would mean "Either this match wasn't struck or else it lit" and "Either this match wasn't struck or else it didn't light" respectively. But since the match was not struck, *both* of these disjunctions are true (because "This match wasn't struck" is true). Therefore, these disjunctions cannot mean the same thing as the counterfactual conditionals, because it is impossible for both of the counterfactual conditionals to be true at the same time. Either "If this match had been struck it would have lit" or "If this match had been struck it wouldn't have lit" is true, but they cannot both be true.

II.5 TRUTH CONDITIONS

We are now able to symbolize the forms of many statements as formulas of the propositional calculus. The next thing is to examine the conditions under which statements that have these forms are true or false. This will allow us to develop precise rules for evaluating the soundness of argument forms symbolized in the propositional calculus and for evaluating the formal analyticity of statement forms symbolized in the propositional calculus.

Formulas of the propositional calculus have an interesting characteristic: if we know whether their atomic parts are true or false, we can determine whether the formulas themselves are true or false.[2]

First let us consider the simplest molecular formulas—those constructed from their atomic parts using a single sentential connective. These are the formulas of the form $\sim P$, $(P \,\&\, Q)$, $(P \vee Q)$, $(P \supset Q)$, and $(P \equiv Q)$. The following table tells us when these formulas are true.

1.	$\sim P$	True if and only if P is false.
2.	$(P \,\&\, Q)$	True if and only if both P and Q are true.
3.	$(P \vee Q)$	True if and only if either P or Q (or both) are true.
4.	$(P \supset Q)$	True if and only if either P is false or Q is true (hence false if and only if P is true and Q is false).
5.	$(P \equiv Q)$	True if and only if either P and Q are both true or P and Q are both false (that is, true if and only if P and Q have the same truth value).

To make this a bit clearer we can construct tables for each of these five formulas telling us exactly when they are true:

[2] Actually, it is not the formulas themselves that are true or false, but rather the statements of that form. So we ought to say that, if we have a statement which has the form of a particular formula, and we know the truth values of the parts of the statement corresponding to the atomic parts of the formula, then we can determine whether the statement itself is true or false.

1.

P	$\sim P$
T	F
F	T

2.

P	Q	$(P \& Q)$
T	T	T
T	F	F
F	T	F
F	F	F

3.

P	Q	$(P \vee Q)$
T	T	T
T	F	T
F	T	T
F	F	F

4.

P	Q	$(P \supset Q)$
T	T	T
T	F	F
F	T	T
F	F	T

5.

P	Q	$(P \equiv Q)$
T	T	T
T	F	F
F	T	F
F	F	T

With the exception of Rule 4, these rules are all rather obvious. For example, $\sim P$ means "It is false that P," and so clearly $\sim P$ is true if and only if P is false. The explanation for Rule 4 turns on the discussion in Section II.4 of the standard use of "if . . . then." Recall that in its standard use, "If P then Q" means the same thing as $(\sim P \vee Q)$. It is this use of "if . . . then" that is symbolized by "$\supset$." So the truth rule for $(P \supset Q)$ should say that $(P \supset Q)$ is true if and only if $(\sim P \vee Q)$ is true. By Rule 3, $(\sim P \vee Q)$ is true if, and only if, either $\sim P$ or Q (or both) are true. And $\sim P$ is true if, and only if, P is false. Therefore, $(\sim P \vee Q)$ is true if, and only if, either P is false or Q is true. If we tabulate this, we get Table 4 for the conditional.

Repeated application of these truth rules will determine the truth or falsity of complex formulas. Suppose we know that P and Q are true, and R and S are false, and we want to determine whether the formula $\{[(P \& \sim S) \supset \sim Q] \equiv \sim (R \vee S)\}$ is true. We begin with the smallest parts. S is false, so $\sim S$ is true. Then P and $\sim S$ are both true, so by Rule 2, $(P \& \sim S)$ is true. Q is true, so by Rule 1, $\sim Q$ is false. Then by Rule 4, as $(P \& \sim S)$ is true and $\sim Q$ is false, the conditional $[(P \& \sim S) \supset \sim Q]$ is false. Neither R nor S is true, so by Rule 3, $(R \vee S)$ is false. Then by Rule 1, $\sim (R \vee S)$ is true. But then the two sides of the biconditional $\{[(P \& \sim S) \supset \sim Q] \equiv \sim (R \vee S)\}$ have different truth values (the left side is false and the right side is true), and so, by Rule 5, the biconditional is false. We can diagram this procedure as follows:

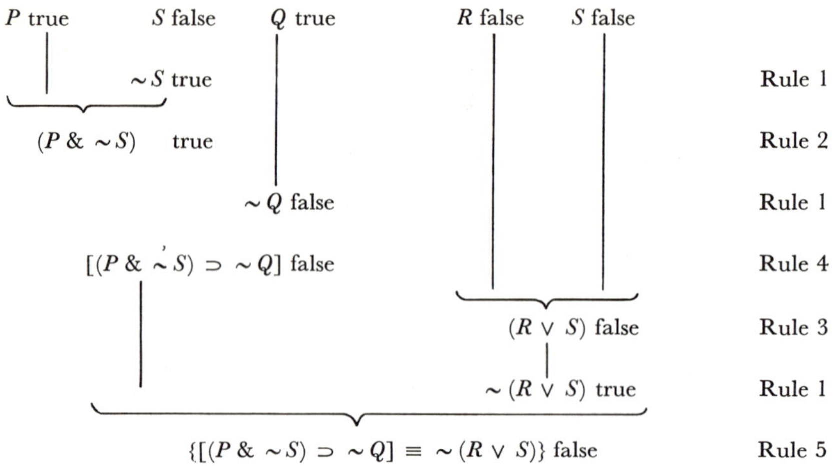

Given the truth values of the atomic parts of a formula, it becomes a purely mechanical matter to calculate the truth value of the whole formula using the above rules of truth. We begin with the smallest parts of the formula, calculate their truth values, and then work out until we get the whole formula. The calculation of the truth value parallels the construction we would use in building the formula from its atomic parts using the rules in Section II.2.

Given a particular formula, $[(P \ \& \sim Q) \equiv (\sim(P \vee Q) \supset Q)]$, we can tabulate the truth values of that formula under all possible combinations of truth values of its atomic parts. For this formula, the following table would be constructed:

Step 1

P	Q	$[(P \ \& \sim Q) \equiv (\sim(P \vee Q) \supset Q)]$
T	T	
T	F	
F	T	
F	F	

Step 2 Truth values for $\sim Q$

P	Q	$[(P \ \& \sim Q) \equiv (\sim(P \vee Q) \supset Q)]$
T	T	F
T	F	T
F	T	F
F	F	T

Step 3 Truth values for $(P \,\&\, \sim Q)$

P	Q	$[(P \,\&$	$\sim Q)$	$\equiv$	$(\sim (P$	$\vee\, Q)$	$\supset Q)]$
T	T	F	F				
T	F	T	T				
F	T	F	F				
F	F	F	T				

Step 4 Truth values for $(P \vee Q)$

P	Q	$[(P \,\&$	$\sim Q)$	$\equiv$	$(\sim (P$	$\vee\, Q)$	$\supset Q)]$
T	T	F	F			T	
T	F	T	T			T	
F	T	F	F			T	
F	F	F	T			F	

Step 5 Truth values for $\sim (P \vee Q)$

P	Q	$[(P \,\&$	$\sim Q)$	$\equiv$	$(\sim (P$	$\vee\, Q)$	$\supset Q)]$
T	T	F	F		F	T	
T	F	T	T		F	T	
F	T	F	F		F	T	
F	F	F	T		T	F	

Step 6 Truth values for $(\sim (P \vee Q) \supset Q)$

P	Q	$[(P \,\&$	$\sim Q)$	$\equiv$	$(\sim (P$	$\vee\, Q)$	$\supset Q)]$
T	T	F	F		F	T	T
T	F	T	T		F	T	T
F	T	F	F		F	T	T
F	F	F	T		T	F	F

Step 7 Truth values for $[(P \,\&\, \sim Q) \equiv (\sim (P \vee Q) \supset Q)]$

P	Q	$[(P \,\&$	$\sim Q)$	$\equiv$	$(\sim (P$	$\vee\, Q)$	$\supset Q)]$
T	T	F	F	F	F	T	T
T	F	T	T	T	F	T	T
F	T	F	F	F	F	T	T
F	F	F	T	T	T	F	F

This table is called a *truth table.* On the left-hand side of a truth table we list all possible combinations of truth values for the atomic parts of the formula. On the right-hand side are columns corresponding to each occurrence of a sentential connective in the formula, and then in that column we list the truth value of the part of the formula that is constructed using that connective. We enclose the column of truth values that are the truth values of the entire formula in a box.

In constructing a truth table, one must be sure to list *all* possible combinations of truth values of its atomic parts on the left. In general, if a formula has n atomic parts, there are 2^n combinations of truth values. Thus if a formula has two atomic parts there are four combinations of truth values, if it has three atomic parts there are eight combinations of truth values, if it has four atomic parts there are sixteen combinations of truth values, and so on. Clearly, if a formula has three or more atomic parts it is easy to omit one or more combinations of truth values inadvertently. To guard against this it is best to list the truth combinations in some fixed order. One such order is the following: We begin with the atomic part we have listed on the extreme right, and then alternate truth values: T, F, T, F, and so on, until we have 2^n truth values listed. Then go to the next atomic part and alternate truth values *two at a time*: T, T, F, F, T, T, and so on. For the next atomic part we alternate four at a time: T, T, T, T, F, F, F, F, T, T, T, T. Then we alternate eight at a time, and so on. In this way we will get all of the combinations of truth values. If we have four atomic parts, P, Q, R, S, the left-hand side of the truth table will look as follows:

P	Q	R	S
T	T	T	T
T	T	T	F
T	T	F	T
T	T	F	F
T	F	T	T
T	F	T	F
T	F	F	T
T	F	F	F
F	T	T	T
F	T	T	F
F	T	F	T
F	T	F	F
F	F	T	T
F	F	T	F
F	F	F	T
F	F	F	F

II.5 EXERCISES

Construct truth tables for the following formulas:

1. $\sim(P \supset \sim\sim P)$
2. $[(P \supset Q) \vee \sim(P \,\&\, \sim Q)]$
3. $[(P \equiv Q) \equiv \sim(P \equiv \sim Q)]$
4. $[((P \,\&\, Q) \supset R) \,\&\, \sim((R \vee P) \equiv \sim(P \supset Q))]$
5. $[\sim(P \supset Q) \supset \sim([(R \vee P) \,\&\, \sim(Q \supset \sim P)]$
   $\equiv \sim[(R \,\&\, \sim Q) \supset (P \,\&\, \sim R)])]$
6. $([(P \vee Q) \supset (R \,\&\, S)]$
   $\equiv [([P \supset R] \,\&\, [P \supset S]) \,\&\, ([Q \supset R] \,\&\, [Q \supset S])])$
7. $[\sim(P \,\&\, Q) \equiv (\sim P \,\&\, \sim Q)]$

II.6 TAUTOLOGIES

Our main interests in logic are the formal analyticity of statements and the soundness of argument forms. Using truth tables we can give very neat characterizations of these concepts for statements and arguments whose forms can be symbolized in the propositional calculus. Let us begin by discussing formal analyticity.

Consider the formally analytic statement "Either it is snowing or it is not snowing." This statement has the form $(P \vee \sim P)$. Any statement that has this form will be analytic. Why is this so? Because regardless of what we let P mean, either P or $\sim P$ will be true, and so $(P \vee \sim P)$ will be true. To make this clearer, consider the truth table for $(P \vee \sim P)$:

P	$(P \vee \sim P)$	
T	[T]	F
F	[T]	T

Notice that $(P \vee \sim P)$ comes out true on every line of this truth table. This is enough to make it formally analytic, because regardless of how we interpret P, it will have one of the truth values listed in the truth table, and then $(P \vee \sim P)$ will be true. Thus we can see by examining the truth table that any statement having the form $(P \vee \sim P)$ will be true, regardless of whether P is true or false, and so we can tell, just by considering the meaning of the statement, that it is true. This means that the statement is analytic. So any statement of the form $(P \vee \sim P)$ is analytic, and hence this statement form is formally analytic.

Now let us generalize the above discussion. A formula that comes out true on every line of its truth table is called a *tautology*. $(P \vee \sim P)$ is therefore a tautology. If some formula, P, is a tautology, we will write $[]\!- P$ (read: "yields P"). For example, we can write $[]\!- (P \vee \sim P)$. Now suppose that we have some formula of that propositional calculus that is a tautology, such as $\{[P \supset (Q \supset R)] \equiv [(P \,\&\, Q) \supset R]\}$. Consider any statement that has this form. Then, even without knowing whether P, Q, and R are true, we can verify that the statement itself is true just by seeing that it comes out true on every line of its truth table. This statement will therefore be true regardless of what truth values P, Q, and R have. Thus any statement of this form will be analytic, and the statement form will be formally analytic. Therefore, we can conclude that *if a formula is a tautology, then every statement of that form is formally analytic.*

The converse of this principle is also true: *if a formula of the propositional calculus is the form of a formally analytic statement, then the formula is a tautology.* To see that this principle is true, suppose we have some formula of the propositional calculus that is the form of a formally analytic statement. Then every statement of that form must be true. Thus the formula must come out true on every line of its truth table, and so it must be a tautology. $[P \supset (P \vee Q)]$ is the form of a formally analytic statement. Every statement of this form is analytic; for example, the statement "If John comes, then either John or Joe will come" is analytic. Now we want to show that because it is formally analytic it is a tautology. Let us consider any assignment of truth values to the atomic parts of $[P \supset (P \vee Q)]$; let P be false and Q be true. Then we can construct some actual statement having the form $[P \supset (P \vee Q)]$ in which what corresponds to P is false, and in which what corresponds to Q is true. We might consider the statement "If Kennedy was the 35th President of the United States, then either Kennedy or Johnson was the 35th President of the United States." This statement is formally analytic, so it is true. Thus $[P \supset (P \vee Q)]$ must come out true on that line of the truth table where P is false and Q is true. By repeating this for all the different combinations of truth values for P and Q we see that $[P \supset (P \vee Q)]$ must come out true on every line of its truth table, therefore it is a tautology.

We can conclude in general then that *a formula of the propositional calculus is a tautology if, and only if, it is the form of a formally analytic statement.* Thus truth tables and tautologies give us a way of talking about the formal analyticity of any statement whose form can be expressed in the propositional calculus.

An important fact about tautologies is that *the conjunction of two tautologies is itself a tautology*. That is, if we begin with two formulas, *P* and *Q*, and ▯– *P* and ▯– *Q*, then ▯– (*P* & *Q*). For suppose that ▯– *P* and ▯– *Q*. Then any assignment of truth values to the atomic parts of *P* and *Q* will make them both true. But then any such assignment of truth values will make the conjunction (*P* & *Q*) true, and so it is also a tautology.

Two concepts that are closely related to that of a tautology are those of a truth-functionally consistent formula and a truth-functionally inconsistent formula. We say that a formula is *truth-functionally consistent* if, and only if, it is true on at least one line of its truth table. A formula is *truth-functionally inconsistent* if, and only if, it is not truth-functionally consistent; that is, if and only if it is false on every line of its truth table. A formula is false on every line of its truth table if, and only if, its negation is true on every line of its truth table. Thus a formula is truth-functionally inconsistent if, and only if, its negation is a tautology.

II.6 EXERCISES

Show that the following statements are true.

1. Every tautology is truth-functionally consistent.
2. It is possible for two formulas to be each truth-functionally consistent while their conjunction is truth-functionally inconsistent (give an example).
3. If one formula is a tautology, and a second formula is truth-functionally consistent, then their conjunction is truth-functionally consistent.
4. It is possible for both a formula and its negation to be truth-functionally consistent.

II.7 TAUTOLOGICAL IMPLICATION AND EQUIVALENCE

Next let us discuss those argument forms that can be expressed in the propositional calculus. Consider the following argument:

If there is an early thaw in the mountains the river will flood.
There will be an early thaw in the mountains.

Therefore, the river will flood.

The form of this argument can be symbolized as follows:

$(M \supset R)$
M

Therefore, R.

We saw in Chapter I that an argument form is sound if, and only if, its corresponding conditional is formally analytic. The corresponding conditional of this argument is $\{[(M \supset R) \& M] \supset R\}$. This is a formula of the propositional calculus, so it is formally analytic if, and only if, it is a tautology. Let us then construct its truth table:

M	R	$\{[(M \supset R)$	$\& M]$	$\supset R\}$
T	T	T	T	**T**
T	F	F	F	**T**
F	T	T	F	**T**
F	F	T	F	**T**

We see that the corresponding conditional is a tautology, so the argument form is sound.

In general, whenever we have an argument form expressed in the propositional calculus, we can assess its soundness by constructing a truth table and determining whether its corresponding conditional is a tautology. This leads to the definition of tautological implication: If we have a finite set of formulas $P_1, \ldots, P_n$, and a single formula Q, we say that the set of formulas $P_1, \ldots, P_n$ *tautologically implies* Q if, and only if, the conditional $[(P_1 \& \ldots \& P_n) \supset Q]$ is a tautology. If $P_1, \ldots, P_n$ tautologically implies Q, we can symbolize this as $P_1, \ldots, P_n \vDash Q$. Then, for example $P, (P \supset Q), R \vDash (Q \& R)$ (that is, the set of formulas P, $(P \supset Q)$, and R tautologically implies $(Q \& R)$) if, and only if, the formula $\{[(P \& (P \supset Q)) \& R] \supset (Q \& R)\}$ is a tautology. If a set of formulas tautologically implies another formula, the latter formula is said to be a *tautological consequent* of the set of formulas.

Our notion of tautological implication is so constructed that if an argument form is expressed in the propositional calculus, it is sound if, and only if, its premises tautologically imply its conclusion. This is the same as saying that its corresponding conditional is a tautology.

There are sound argument forms whose premises do not tautologically imply their conclusions, but these are also argument forms that cannot be expressed in the propositional calculus. The following argument form is sound, but it cannot be expressed in the propositional calculus:

All *A* are *B*.
All *B* are *C*.

Therefore, all *A* are *C*.

The propositional calculus can only deal with those argument forms that can be built up using only the sentential connectives.

The simplest case of tautological implication is where a single formula implies another formula. Thus, P tautologically implies $(P \vee Q)$. Notice that given two formulas, P and Q, P tautologically implies Q if, and only if, the conditional $(P \supset Q)$ is a tautology. This is the same as saying that every line of the truth table on which P is true is also a line on which Q is true. Roughly speaking, to say that one formula tautologically implies another simply means that whenever the first is true, so is the second.

It is useful to have a way of distinguishing between the formula that does the implying and the formula that is implied in a tautological implication. In analogy with conditionals we can say that the first is the *antecedent* of the implication, and the second is the *consequent*.

The following tautological implications are important and are easily verified using truth tables:

I1.[3] $(P \& Q) \models P$ } (simplification)
I2. $(P \& Q) \models Q$ }
I3. $P \models (P \vee Q)$ } (addition)
I4. $P \models (Q \vee P)$ }
I5. $\sim P \models (P \supset Q)$
I6. $Q \models (P \supset Q)$
I7. $\sim (P \supset Q) \models P$
I8. $\sim (P \supset Q) \models \sim Q$
I9. $P, (P \supset Q) \models Q$ (*modus ponens*)

[3] I stands for implication, I1 and so forth.

I10. $\sim Q, (P \supset Q) \;\square\!- \sim P$ (*modus tollens*)
I11. $\sim P, (P \vee Q) \;\square\!- Q$ (disjunctive syllogism)
I12. $(P \supset Q), (Q \supset R) \;\square\!- (P \supset R)$ (hypothetical syllogism)
I13. $P, Q \;\square\!- (P \;\&\; Q)$ (adjunction)
I14. $(P \vee Q), (P \supset R), (Q \supset R) \;\square\!- R$ (dilemma)

We can generalize I13 somewhat. Given any number, say n, of formulas $P_1, \ldots, P_n$, we have:

I15. $P_1, \ldots, P_n \;\square\!- (P_1 \;\&\; \ldots \;\&\; P_n)$

Notice that in general, $P_1, \ldots, P_n \;\square\!- Q$ if, and only if,

$$(P_1 \;\&\; \ldots \;\&\; P_n) \;\square\!- Q,$$

because both of these hold if, and only if, $\square\!- [(P_1 \;\&\; \ldots \;\&\; P_n) \supset Q]$.

Now let us examine each of the above implications and see why it holds:

I1 and I2 both hold because a conjunction is true if and only if both conjuncts are true.
I3 and I4 hold because a disjunction is true if either disjunct is true.
I5 holds because a conditional is true if its antecedent is false.
I6 holds because a conditional is true if its consequent is true.
I7 and I8 hold because a conditional is only false when its antecedent is true and its consequent false.
19: Suppose P is true and $(P \supset Q)$ is true. Then we have a true conditional having a true antecedent. But then, in order for the conditional to be true, the consequent must also be true. Thus Q must be true.
I10: Suppose $\sim Q$ is true and $(P \supset Q)$ is true. Then we have a true conditional having a false consequent. But then, in order for the conditional to be true, the antecedent must be false. So P must be false; that is, $\sim P$ must be true.
I11: Suppose $\sim P$ and $(P \vee Q)$ are both true. In order for the disjunction to be true, at least one of its disjuncts must be true. But the first disjunct is false, because $\sim P$ is true. Thus the second disjunct, Q, must be true.
I12: This principle says that if we have both "if P then Q" and "if Q then R," then we can conclude that "if P then R." This should be intuitively obvious, but if not it is easily verified with truth tables.
I13: If both P and Q are true, then both conjuncts of $(P \;\&\; Q)$ are true, so the conjunction is true.
I14: Suppose $(P \vee Q)$, $(P \supset R)$, and $(Q \supset R)$ are all true. Since the

disjunction $(P \vee Q)$ is true, either P is true or Q is true. But if P is true then R is true (by the first conditional), and if Q is true then R is true (by the second conditional). Thus if either P or Q is true, then R is true. But we know that either P is true or Q is true, so R is true.
I15: If all of $P_1, \ldots, P_n$ are true, then their conjunction is true.

A concept closely related to tautological implication is that of tautological equivalence. We say that two formulas are *tautologically equivalent* if, and only if, each formula tautologically implies the other. Thus, for example, $(P \& Q)$ is tautologically equivalent to $(Q \& P)$ because $(P \& Q)$ tautologically implies $(Q \& P)$, and $(Q \& P)$ tautologically implies $(P \& Q)$, which can be verified with a truth table. Given two formulas, P and Q, let us abbreviate "P is tautologically equivalent to Q" as "P eq. Q."

Just as tautological implication is the tautologicity of the conditional, tautological equivalence is the tautologicity of the biconditional. That is, given two formulas, P and Q, they are tautologically equivalent if, and only if, the biconditional $(P \equiv Q)$ is a tautology; for example, $\sim\sim P$ is tautologically equivalent to P, because $[]\!- (\sim\sim P \equiv P)$. This is for the following reason: $(P \equiv Q)$ is a tautology if, and only if, P and Q are true on exactly the same lines of the truth table. But this is the same as saying that every line on which P is true is also a line on which Q is true, and every line on which Q is true is also a line on which P is true; that is, P tautologically implies Q, and Q tautologically implies P.

The following tautological equivalences can all be verified very easily using truth tables:

E1.[4] $\sim\sim P$ eq. P (double negation)
E2. $\sim(P \vee Q)$ eq. $(\sim P \& \sim Q)$ } (De Morgan's Laws)
E3. $\sim(P \& Q)$ eq. $(\sim P \vee \sim Q)$ }
E4. $(P \supset Q)$ eq. $(\sim P \vee Q)$
E5. $\sim(P \supset Q)$ eq. $(P \& \sim Q)$
E6. $\sim(P \equiv Q)$ eq. $(P \equiv \sim Q)$
E7. $(P \equiv Q)$ eq. $[(P \supset Q) \& (Q \supset P)]$
E8. $(P \equiv Q)$ eq. $[(P \& Q) \vee (\sim P \& \sim Q)]$
E9. $(P \& (Q \& R))$ eq. $((P \& Q) \& R)$ } (commutative laws)
E10. $(P \vee (Q \vee R))$ eq. $((P \vee Q) \vee R)$ }
E11. $(P \& Q)$ eq. $(Q \& P)$ } (associative laws)
E12. $(P \vee Q)$ eq. $(Q \vee P)$ }

[4] E stands for equivalence, E1 and so forth.

E13. $[P \,\&\, (Q \vee R)]$ eq. $[(P \,\&\, Q) \vee (P \,\&\, R)]$ } (distributive laws)
E14. $[P \vee (Q \,\&\, R)]$ eq. $[(P \vee Q) \,\&\, (P \vee R)]$ }
E15. $[P \supset (Q \supset R)]$ eq. $[(P \,\&\, Q) \supset R]$ (exportation-importation)
E16. $(P \supset Q)$ eq. $(\sim Q \supset \sim P)$ (contraposition)
E17. $[R \vee (P \,\&\, \sim P)]$ eq. R
E18. $(P \vee P)$ eq. P } (idempotency laws)
E19. $(P \,\&\, P)$ eq. P }
E20. $[R \,\&\, (P \vee \sim P)]$ eq. R
E21. $[R \,\&\, (P \,\&\, \sim P)]$ eq. $(P \,\&\, \sim P)$

Let us examine each of these equivalences separately, and see why it is true:

E1. $\sim\sim P$ is true if, and only if, $\sim P$ is not true, and $\sim P$ is not true if, and only if, it is not the case that P is not true; that is, if and only if P is true.

E2. $\sim(P \vee Q)$ is true if, and only if, $(P \vee Q)$ is false. But $(P \vee Q)$ is false if, and only if, both disjuncts are false; that is, if and only if both P and Q are false. But this is the same as saying that $\sim P$ and $\sim Q$ are both true; that is, that $(\sim P \,\&\, \sim Q)$ is true.

E3. $\sim(P \,\&\, Q)$ is true if, and only if, $P \,\&\, Q$ is false. But $(P \,\&\, Q)$ is false if, and only if, at least one conjunct is false; that is, if and only if either P is false or Q is false. But this is the same as saying that either $\sim P$ is true or $\sim Q$ is true; that is, that $(\sim P \vee \sim Q)$ is true.

E4. We saw in Section 4 that $(P \supset Q)$ is true if, and only if, $(\sim P \vee Q)$ is true.

E5. $\sim(P \supset Q)$ is true if, and only if, $(P \supset Q)$ is false. $(P \supset Q)$ is false if, and only if, the antecedent is true and the consequent false, that is, P and $\sim Q$ are true. Thus $(P \supset Q)$ is false if, and only if, $(P \,\&\, \sim Q)$ is true. Thus $\sim(P \supset Q)$ is true if, and only if, $(P \,\&\, \sim Q)$ is true.

E6. $\sim(P \equiv Q)$ is true if, and only if, P and Q have opposite truth values; that is, if and only if P and $\sim Q$ have the same truth values; that is, if and only if $(P \equiv \sim Q)$ is true.

E7. $(P \equiv Q)$ is true if, and only if, P and Q have the same truth value. Thus if P is true then Q must be true, and if Q is true then P must be true. So if $(P \equiv Q)$ is true, then

$$[(P \supset Q) \,\&\, (Q \supset P)]$$

must be true. Conversely, if $[(P \supset Q) \& (Q \supset P)]$ is true, then if P is true, so must Q be true, and if P is false, so must Q be false. Thus $(P \equiv Q)$ is true. Therefore, $(P \equiv Q)$ is true if, and only if, $[(P \supset Q) \& (Q \supset P)]$ is true.

E8. $(P \equiv Q)$ is true if, and only if, P and Q have the same truth value. This is the same as saying that either P and Q are both true; that is, $(P \& Q)$ is true, or P and Q are both false; that is, $(\sim P \& \sim Q)$ is true. Thus $(P \equiv Q)$ is true if, and only if, $[(P \& Q) \vee (\sim P \& \sim Q)]$ is true.

E9 is obvious.

E10 is obvious.

E11 is obvious.

E12 is obvious.

E13. To say that $[P \& (Q \vee R)]$ is true is to say that both P and $(Q \vee R)$ are true. Thus P must be true, and either Q or R must be true. So either $(P \& Q)$ will be true, or $(P \& R)$ will be true. So if $[P \& (Q \vee R)]$ is true, then $[(P \& Q) \vee (P \& R)]$ is true.

Conversely, if $[(P \& Q) \vee (P \& R)]$ is true, then either $(P \& Q)$ or $(P \& R)$ must be true. In either case, P must be true. In addition, in order to have one of $(P \& Q)$ or $(P \& R)$ true, we must have either Q true or R true. So $(Q \vee R)$ must be true. Thus both P and $(Q \vee R)$ must be true, so $[P \& (Q \vee R)]$ must be true. Thus $[P \& (Q \vee R)]$ is true if, and only if, $[(P \& Q) \vee (P \& R)]$ is true.

E14. Verify this with a truth table.

E15. To say that $[P \supset (Q \supset R)]$ is true is to say that if P is true, then $(Q \supset R)$ is true; that is, that if P is true, then if Q is also true, then R is true. Thus if P and Q are both true, then R must be true; that is, $[(P \& Q) \supset R]$ is true. Conversely, if $[(P \& Q) \supset R]$ is true, then if P is true, then if Q is also true, then R will be true; that is, if P is true, then $(Q \supset R)$ will be true; that is, $[P \supset (Q \supset R)]$ is true.

E16. If $(P \supset Q)$ is true, then if Q is false, P must also be false, because if P were true, then the conditional would have a true antecedent and false consequent, which would make it false. Thus if $\sim Q$ is true, then $\sim P$ is true; that is, $(\sim Q \supset \sim P)$ is true.

Conversely, if $(\sim Q \supset \sim P)$ is true, then by the same argument, $(\sim\sim P \supset \sim\sim Q)$ must be true. But $\sim\sim P$ is equivalent to P, and $\sim\sim Q$ is equivalent to Q, so $(P \supset Q)$ must be true.

E17. $[R \vee (P \,\&\, \sim P)]$ is a disjunction, the second disjunct of which can never be true. Thus it is true if, and only if, the first disjunct is true.

E18 is obvious.

E19 is obvious.

E20. $[R \,\&\, (P \vee \sim P)]$ is true if, and only if, both R and $(P \vee \sim P)$ are true. But $(P \vee \sim P)$ is always true. Thus $[R \,\&\, (P \vee \sim P)]$ is true if and only if R is true.

E21. $[R \,\&\, (P \,\&\, \sim P)]$ is true if, and only if, both R and $(P \,\&\, \sim P)$ are true. But $(P \,\&\, \sim P)$ is never true. Thus $[R \,\&\, (P \,\&\, \sim P)]$ is never true. Therefore, $[R \,\&\, (P \,\&\, \sim P)]$ is true when, and only when, $(P \,\&\, \sim P)$ is true, which is never.

The student should commit all of the above tautological implications and equivalences to memory. They should all be fairly obvious when you reflect upon the truth tables for the relevant connectives.

By virtue of E9 and E10, we can simplify the way in which we write some of the formulas of the propositional calculus. If we have a long conjunction, such as $(((P \,\&\, (Q \,\&\, R)) \,\&\, (S \,\&\, T)) \,\&\, U)$, it really makes no difference where we put the inner parentheses. This is tautologically equivalent to $(P \,\&\, (Q \,\&\, (R \,\&\, (S \,\&\, (T \,\&\, U)))))$, and to $(((((P \,\&\, Q) \,\&\, R) \,\&\, S) \,\&\, T) \,\&\, U)$, and to

$$(((P \,\&\, Q) \,\&\, (R \,\&\, S)) \,\&\, (T \,\&\, U)),$$

and so on. So instead of writing formulas with complicated arrangements of parentheses, we can write $(P \,\&\, Q \,\&\, R \,\&\, S \,\&\, T \,\&\, U)$, omitting all but the outermost parentheses. This is the convention we will adopt from now on. We can do the same thing for long disjunctions. Thus we can write such things as $[(P \supset Q) \,\&\, (R \supset S) \,\&\, \sim T]$ and as $[(T \equiv U) \vee \sim R \vee \sim S \vee (R \supset T)]$.

There are several important facts that should be observed about tautological implication and tautological equivalence. First, *if a formula is either tautologically equivalent to, or tautologically implied by a tautology, then it is itself a tautology*. This is because, if either of these is the case, then the formula in question must be true on every line of a truth table on which the tautology is true. Since the tautology is true on every line of the truth table, the formula itself must also be true on every line of its truth table; thus it is a tautology.

Next it should be observed that tautological implication is *transitive*: if one formula (or set of formulas) tautologically implies a second formula, and the second formula tautologically implies a third, then the first formula (or set of formulas) tautologically implies the third. Suppose we have a set of formulas $P_1, \ldots, P_n$ and Q and R, and suppose $P_1, \ldots, P_n \;\square\!- Q$ and $Q \;\square\!- R$. Let us show that it follows from this that $P_1, \ldots, P_n \;\square\!- R$. By the definition of tautological implication, as $P_1, \ldots, P_n \;\square\!- Q$ and $Q \;\square\!- R$, we must have

$$\square\!- [(P_1 \;\&\; \ldots \;\&\; P_n) \supset Q]$$

and $\square\!- (Q \supset R)$. We saw in the last section that the conjunction of two tautologies is a tautology, so

$$\square\!- [((P_1 \;\&\; \ldots \;\&\; P_n) \supset Q) \;\&\; (Q \supset R)].$$

But by I12,

$$[((P_1 \;\&\; \ldots \;\&\; P_n) \supset Q) \;\&\; (Q \supset R)] \;\square\!- [(P_1 \;\&\; \ldots \;\&\; P_n) \supset R].$$

We just saw that if a formula is implied by a tautology then it is itself a tautology, so $\square\!- [(P_1 \;\&\; \ldots \;\&\; P_n) \supset R]$. But then by the definition of tautological implication, $P_1, \ldots, P_n \;\square\!- R$.

Because tautological implication is transitive, we can establish tautological implications using several steps. For example, suppose we want to show that $(P \;\&\; Q) \;\square\!- (P \vee Q)$. We could do this by using a truth table and showing that $\square\!- [(P \;\&\; Q) \supset (P \vee Q)]$. But it is much easier to do it by I1—$(P \;\&\; Q) \;\square\!- P$ or by I3—$P \;\square\!- (P \vee Q)$. Therefore, by the transitivity of implication, $(P \;\&\; Q) \;\square\!- (P \vee Q)$. Using the transitivity of tautological implication in this way we can string simple implications together to establish more complicated implications. If we want to show that one formula, P, implies another formula, Q, we might do this by showing that P implies some other formula, and that the second formula implies a third formula (and hence P implies the third formula); then the third formula implies a fourth formula (and hence P implies the fourth formula), and so on until we obtain the formula Q. In other words, if we have a string of implications of the form $P \;\square\!- P_1$, $P_1 \;\square\!- P_2$, $P_2 \;\square\!- P_3, \ldots, P_n \;\square\!- Q$, it follows that $P \;\square\!- Q$. This is because, as $P \;\square\!- P_1$ and $P_1 \;\square\!- P_2$, we must have $P \;\square\!- P_2$, and then as $P_2 \;\square\!- P_3$, we must have $P \;\square\!- P_3$, and so on until we establish that $P \;\square\!- Q$.

Notice that in constructing strings of implications like this, we can also use tautological equivalences, because if two formulas are tautologically equivalent, then each tautologically implies the other. If we want to show that $(P \equiv Q) \models (\sim P \supset \sim Q)$, we may use E7 to observe that $(P \equiv Q) \models [(P \supset Q) \& (Q \supset P)]$. Then by I2,

$$[(P \supset Q) \& (Q \supset P)] \models (Q \supset P)$$

and by E16, $(Q \supset P) \models (\sim P \supset \sim Q)$. Finally, by the transitivity of implication, $(P \equiv Q) \models (\sim P \supset \sim Q)$.

Given that tautological implication is transitive, it is easy to see that tautological equivalence is also transitive: if one formula is tautologically equivalent to a second, and the second is tautologically equivalent to a third, then the first is tautologically equvialent to the third. Suppose that P eq. Q and Q eq. R; then $P \models Q$ and $Q \models R$, and so by the transitivity of tautological implication, $P \models R$. Also $Q \models P$ and $R \models Q$, and so once again by the transitivity of tautological implication, $R \models P$. Then because both $P \models R$ and $R \models P$, we can conclude that P eq. R.

A further important fact about tautological implication is that it is *adjunctive*: given any three formulas, P, Q, and R, if $P \models Q$ and $P \models R$, then $P \models (Q \& R)$. In other words, if a formula implies each of two other formulas, then it implies their conjunction. To see that this is true, suppose $P \models Q$ and $P \models R$. Any assignment of truth values to the atomic parts of P, Q, and R that makes P true will make both Q and R true, so any such assignment will make $(Q \& R)$ true, and hence $P \models (Q \& R)$. To illustrate this, suppose we want to show that P, $(P \equiv Q) \models Q$. Clearly $P, (P \equiv Q) \models P$ and $P, (P \equiv Q) \models (P \equiv Q)$. By E7, $(P \equiv Q) \models [(P \supset Q) \& (Q \supset P)]$, and by I1,

$$[(P \supset Q) \& (Q \supset P)] \models (P \supset Q).$$

Thus by the transitivity of implication, $P, (P \equiv Q) \models (P \supset Q)$. Then by the fact that implication is adjunctive, $P, (P \equiv Q) \models [P \& (P \supset Q)]$. By I9, $[P \& (P \supset Q)] \models Q$. Thus by transitivity again, $P, (P \equiv Q) \models Q$.

We can generalize the adjunctivity of tautological implication somewhat: if a set of formulas $P_1, \ldots, P_n$ tautologically implies each formula in another set $Q_1, \ldots, Q_m$, then $P_1, \ldots, P_n$ implies the conjunction $(Q_1 \& \ldots \& Q_m)$. Suppose the set $P_1, \ldots, P_n$ implies each of $Q_1, \ldots, Q_m$. Then as $P_1, \ldots, P_n \models Q_1$ and $P_1, \ldots, P_n \models Q_2$, we must have $P_1, \ldots, P_n \models (Q_1 \& Q_2)$. Then as $P_1, \ldots, P_n \models Q_3$ we must have $P_1, \ldots, P_n \models (Q_1 \& Q_2 \& Q_3)$. And so on. Thus $P_1, \ldots, P_n \models (Q_1 \& \ldots \& Q_n)$.

We can also generalize the transitivity of tautological implication somewhat: if each formula in a set $Q_1, \ldots, Q_m$ of formulas is tautologically implied by the set of formulas $P_1, \ldots, P_n$ and $Q_1, \ldots, Q_m \Box\!- R$, then $P_1, \ldots, P_n \Box\!- R$. Suppose each of $Q_1, \ldots, Q_m$ are implied by $P_1, \ldots, P_n$. Then as we have just seen, $P_1, \ldots, P_n \Box\!- (Q_1 \,\&\ldots\&\, Q_m)$. If $Q_1, \ldots, Q_m \Box\!- R$, then $(Q_1 \,\&\ldots\&\, Q_m) \Box\!- R$. So by transitivity, $P_1, \ldots, P_n \Box\!- R$. Let us call this the *strong transitivity* of tautological implication. The difference between transitivity and strong transitivity is that in the former we have a set of formulas implying a *single* formula, which in turn implies another formula, whereas in the latter we have a set of formulas implying a *second set* of formulas, which in turn implies another formula.

Another important fact about implication is the following: if we have a set of formulas, say $R_1, \ldots, R_n, P$, which tautologically implies some formula Q, then $R_1, \ldots, R_n \Box\!- (P \supset Q)$. This is called the rule of *conditionalization.* It is true for the following reason. Suppose $R_1, \ldots, R_n, P \Box\!- Q$. By the definition of tautological implication, this means that the conditional $[(R_1 \,\&\ldots\&\, R_n \,\&\, P) \supset Q]$ is a tautology. By E15 this is tautologically equivalent to the conditional

$$[(R_1 \,\&\ldots\&\, R_n) \supset (P \supset Q)].$$

We have seen that a formula tautologically equivalent to a tautology is itself a tautology, so $\Box\!- [(R_1 \,\&\ldots\&\, R_n) \supset (P \supset Q)]$. Then $R_1, \ldots, R_n \Box\!- (P \supset Q)$.

The rule of conditionalization is of considerable use in establishing tautological implications; for example, suppose we want to show that $P \Box\!- [Q \supset (P \,\&\, Q)]$. By I13, $P, Q \Box\!- (P \,\&\, Q)$. Then by the rule of conditionalization, $P \Box\!- [Q \supset (P \,\&\, Q)]$. Or again, suppose we want to show that $(P \supset R) \Box\!- [(P \vee Q) \supset [(Q \supset R) \supset R]]$. By I14, $(P \vee Q), (P \supset R), (Q \supset R) \Box\!- R$. By the rule of conditionalization, $(P \vee Q), (P \supset R) \Box\!- [(Q \supset R) \supset R]$. Then by a second application of the rule of conditionalization,

$$(P \supset R) \Box\!- [(P \vee Q) \supset [(Q \supset R) \supset R]].$$

Using these facts about tautological implication, it becomes much easier to show that one formula or set of formulas implies another formula. For example, suppose we want to show that $(\sim P \vee Q), (\sim Q \vee R) \Box\!- (\sim P \vee R)$. We might reason as follows. Clearly $(\sim P \vee Q), (\sim Q \vee R) \Box\!- (\sim P \vee Q)$ and $(\sim P \vee Q), (\sim Q \vee R) \Box\!- (\sim Q \vee R)$. By E4, $(\sim P \vee Q) \Box\!- (P \supset Q)$, and $(\sim Q \vee R) \Box\!- (Q \supset R)$. So by transitivity, $(\sim P \vee Q), (\sim Q \vee R) \Box\!- (P \supset Q)$ and $(\sim P \vee Q)$,

$(\sim Q \vee R) \models (Q \supset R)$. By I12, $(P \supset Q), (Q \supset R) \models (P \supset R)$, and by E4, $(P \supset R) \models (\sim P \vee R)$. Then by strong transitivity, $(\sim P \vee Q)$, $(\sim Q \vee R) \models (\sim P \vee R)$.

II.7 EXERCISES

A*. Using the transitivity, strong transitivity, and adjunctivity of tautological implication, together with the rule of conditionalization, the tautological equivalences E1–E21, the tautological implications I1–I15, and the fact that a set of formulas tautologically implies each formula in that set, establish the following tautological implications:

1. $P \models [Q \supset (R \supset [P \,\&\, (Q \,\&\, R)])]$
2. $(P \,\&\, Q) \models [(P \vee R) \,\&\, Q]$
3. $(P \,\&\, Q) \models [(P \vee R) \,\&\, (Q \vee S)]$
4. $(P \equiv Q), \sim Q \models \sim P$
5. $(P \supset Q) \models [(Q \supset R) \supset (\sim R \supset \sim P)]$

B. Show that the following statements are true:

1. A truth-functionally inconsistent formula tautologically implies every formula.
2. A tautology is tautologically implied by every formula.
3. Any two tautologies are tautologically equivalent.

II.8 SUBSTITUTION INSTANCES

In the previous section a number of tautological implications and tautological equivalences were listed. All of these can be generalized; for example, according to I2, $(P \,\&\, Q)$ tautologically implies Q. This says that a particular formula, $(P \,\&\, Q)$, which is the conjunction of the two atomic formulas P and Q, implies its second conjunct. But in fact, *any* conjunction tautologically implies its second conjunct. In other words, I2 can be regarded as an implication *scheme*—given any conjunction, it implies its second conjunct. If we substitute any formulas at all for P and Q, we will obtain another tautological implication; for example, if we substitute $(R \supset S)$ for P and $\sim T$ for Q, we see that $[(R \supset S) \,\&\, \sim T]$ tautologically implies $\sim T$.

* The more difficult exercises are marked with asterisks. See the Preface.

For each of the implications and equivalences listed in Section II.7, it is easily seen that if we substitute other formulas for the various atomic parts of the formulas listed there, we will obtain new implications and equivalences.[5] According to E6, for example, $\sim(P \supset Q)$ is equivalent to $(P \,\&\, \sim Q)$. This is because a conditional is false if, and only if, its antecedent is true and its consequent false. Substituting $(S \equiv \sim T)$ for P and $[(R \,\&\, S) \supset \sim S]$ for Q, we see that

$$\sim[(S \equiv \sim T) \supset [(R \,\&\, S) \supset \sim S)]]$$

is equivalent to $[(S \equiv \sim T) \,\&\, \sim[(R \,\&\, S) \supset \sim S]]$.

The above considerations lead us to define the general concept of a *substitution instance*. We say that one formula is a substitution instance of a second formula when the first can be obtained from the second by substituting formulas for some of the atomic parts of the second formula in such a way that the same formula is substituted for the same atomic part each time it occurs. In the above cases, what we have done is to construct substitution instances of the formulas in the list of tautological implications and tautological equivalences. The principal interest of substitution instances is that any substitution instance of a tautology will be another tautology; analogously, if we make the same substitutions in both the antecedent and consequent of an implication, we will obtain another implication, which is what we did above.

Now consider some examples of substitution instances.

$$[(R \equiv S) \supset (T \,\&\, (R \equiv S))]$$

is a substitution instance of $[P \supset (T \,\&\, P)]$, because the former results from substituting $(R = S)$ for each occurrence of P in $[P \supset (T \,\&\, P)]$. On the other hand, the formulas $[(R \equiv S) \supset (T \,\&\, P)]$ and $[(R \equiv S) \supset (T \,\&\, Q)]$ are *not* substitution instances of $[P \supset (T \,\&\, P)]$ because they do not result from substituting the same formula for *every* occurrence of P. In constructing a substitution instance, if you substitute a formula for one occurrence of an atomic part, you must substitute the same formula for every other occurrence of that same atomic part. If we did not make this restriction it would not be true that every substitution instance of a tautology is a tautology. For example, $(P \vee \sim P)$ is a tautology. But if Q is substituted for the

[5] In fact, we already made covert use of this fact several times in the last section when we discussed the transitivity, adjunctivity, and so on of tautological implication.

first occurrence of P but not for the second, the result is $(Q \vee \sim P)$, which is not a tautology.

Although we must substitute the same formula for the same atomic part each time it occurs, it is not necessary, in constructing a substitution instance, to substitute different formulas for different atomic parts. $[(S \,\&\, \sim T) \supset \sim(S \,\&\, \sim T)]$, for example, is a substitution instance of $(P \supset \sim Q)$. This results from substituting $(S \,\&\, \sim T)$ for every occurrence of P *and* every occurrence of Q. Or, if we begin with the tautology $[(P \,\&\, Q) \supset P]$ and substitute R for each occurrence of P and also for each occurrence of Q, we get $[(R \,\&\, R) \supset R]$, which is also a tautology.

Notice also that in constructing substitution instances we can use atomic parts that already occurred in the original formula. To take a simple case, $(Q \supset Q)$ is a substitution instance of $(P \supset Q)$. A bit more complicated case is: $(Q \supset R)$ is a substitution instance of $(P \supset Q)$. The latter results from substituting Q for every occurrence of P and R for every occurrence of Q in $(P \supset Q)$. Another example: $(Q \supset P)$ is a substitution instance of $(P \supset Q)$. In the latter example, we substitute Q for each occurrence of P, and P for each occurrence of Q.

It must be emphasized that in constructing substitution instances of a formula, substitutions can only be made for the *atomic* parts, never for the molecular parts. Therefore, $(P \supset Q)$, which results from substituting Q for $\sim R$ in $(P \supset \sim R)$, is *not* a substitution instance of $(P \supset \sim R)$. On the other hand, $(P \supset \sim\sim Q)$, which results from substituting $\sim Q$ for R in $(P \supset \sim R)$, *is* a substitution instance. This restriction on substitutions is essential if we are to have all substitution instances of tautologies be tautologies. If we were allowed to substitute for molecular parts, this could destroy some of the structure of a tautology and result in a formula that is not a tautology. For example, $[Q \supset (P \vee \sim P)]$ is a tautology, because it is a conditional whose consequent is always true. But if we were allowed to substitute something else for the molecular part $(P \vee \sim P)$, such as $(S \equiv T)$, giving us $[Q \supset (S \equiv T)]$, we would destroy the structure that made the original formula a tautology, and end up with a formula that is not a tautology.

It should be noted that substitution is a purely structural concept. It has nothing to do with the meaning of a formula—only with the way it is constructed. Because of this, $[(P \supset \sim Q) \,\&\, (R \equiv S)]$ is a substitution instance of $[T \,\&\, (R \equiv S)]$, but $[(R \equiv S) \,\&\, (P \supset \sim Q)]$ is not, even though they are tautologically equivalent.

So far we have only talked about individual formulas being substitution instances of individual formulas. But sometimes we want to talk also about *sets* of formulas being substitution instances of sets of formulas. Given two sets of formulas, we say that the first is a substitution instance of the second f, and only if, the first set of formulas is the set of substitution instances of formulas in the second set, which are obtained by making the same substitutions for the same atomic parts in all of those formulas. Given the set of formulas $(P \supset Q)$, $(P \equiv \sim Q)$, $\sim R$ the set of formulas $[(S \& T) \supset \sim U]$, $[(S \& T) \equiv \sim\sim U]$, $\sim(S \equiv (U \& \sim T))$, is the substitution instance of the first set, which results from substituting $(S \& T)$ for P, $\sim U$ for Q, and $(S \equiv (U \& \sim T))$ for R. The only thing you must watch in constructing a substitution instance of a set of formulas is that you make the same substitutions in *ail* of the formulas in the set.

What is to be established now is that if we begin with a tautology and construct a substitution instance of it, the result will be another tautology; that is, *every substitution instance of a tautology is a tautology*. Consider an example. $(P \vee \sim P)$ is a tautology, and no matter what we substitute for P we get another tautology; for example, if we substitute $(R \supset S)$ for P, giving us $[(R \supset S) \vee \sim(R \supset S)]$, this is also a tautology. This must be a tautology because every assignment of truth values to R and S will give $(R \supset S)$ one of the truth values of P represented in the truth table for $(P \vee \sim P)$, and as $(P \vee \sim P)$ is true on that line of its truth table, $[(R \supset S) \vee \sim(R \supset S)]$ must also be true for that value of $(R \supset S)$.

In general, suppose we have two formulas, P and Q, where P is a substitution instance of Q, and Q is a tautology. Then every assignment of truth values to the atomic parts of Q will make Q true. Now consider P. Consider any assignment of truth values to the atomic parts of P. Then consider the truth values of those parts of P that were substituted for atomic parts of Q. Whatever the truth values of those parts, they are the same as the combination of truth values for the corresponding atomic parts of Q on some line of the truth table for Q, and as Q came out true for that assignment of truth values, P must also come out true. Thus P must also be true on every line of its truth table. This can be illustrated with a concrete case. $[(P \& Q) \supset Q]$ is a tautology. $[([R \supset \sim S] \& Q) \supset Q]$ is a substitution instance, which results from substituting $(R \supset \sim S)$ for P. Consider the truth tables for both formulas:

P	Q	$[(P \& Q) \supset Q]$	
T	T	T	T
T	F	F	T
F	T	F	T
F	F	F	T

R	S	Q	$[([R \supset {\sim}S] \& Q) \supset Q]$			
T	T	T	F	F	F	T
T	T	F	F	F	F	T
T	F	T	T	T	T	T
T	F	F	T	T	F	T
F	T	T	T	F	T	T
F	T	F	T	F	F	T
F	F	T	T	T	T	T
F	F	F	T	T	F	T

Each line of the latter truth table corresponds to a line of the former truth table in which Q has the same value in both, and $(R \supset {\sim}S)$ has the same value as P. The correspondence is indicated with arrows. The truth value of a formula is determined by the truth values of its parts. So if Q has the same value in both formulas, and $(R \supset {\sim}S)$ has the same value as P, then $[([R \supset {\sim}S] \& Q) \supset Q]$ must have the same value as $[(P \& Q) \supset Q]$.

Look at the first line of the truth table for $[([R \supset {\sim}S] \& Q) \supset Q]$; $[R \supset {\sim}S]$ is false, and Q is true. This corresponds to the line on which P is false and Q is true. Now calculate the values of the whole formulas on those lines of their truth tables. The calculations are exactly the same:

$[R \supset {\sim}S]$ false Q True	P false Q True
$([R \supset {\sim}S] \& Q)$ False	$(P \& Q)$ False
$[([R \supset {\sim}S] \& Q) \supset Q]$ True	$[(P \& Q) \supset Q]$ True

The steps of calculation and the starting points are the same, the end results must therefore be the same. Thus on each line of its truth table, $[([R \supset {\sim}S] \& Q) \supset Q]$ will have the same value as $[(P \& Q) \supset Q]$ on the corresponding line of its truth table. Therefore, because the latter is true on every line of its truth table, so must the former be. Consequently, $[([R \supset {\sim}S] \& Q) \supset Q]$ is also a tautology.

The fact that substitution instances of tautologies are themselves tautologies will often make it much easier to establish that a complex formula is a tautology. Suppose we are asked whether the following formula is a tautology:

$$\{[R \equiv (S \,\&\, [(\{[(P \vee \sim Q) \supset T] \equiv U\} \,\&\, \sim V) \vee W])] \\ \vee \sim[R \equiv (S \,\&\, [(\{[(P \vee \sim Q) \supset T] \equiv U\} \,\&\, \sim V) \vee W])]\}$$

It has eight atomic parts, so if we were to construct its truth table it would be 256 lines long. Clearly this would take a long time to work out. On the other hand, if we note that it is a substitution instance of $(P \vee \sim P)$, which we already know to be a tautology, we could say immediately that this formula is also a tautology.

Next we want to show that, with the help of substitution, the tautological implications and equivalences listed in Section II.7 can be thought of as implication and equivalence *schemes* rather than particular implications and equivalences. This means the following: According to I2, $(P \,\&\, Q)$ tautologically implies P, which is the same as saying that the conditional $[(P \,\&\, Q) \supset P]$ is a tautology. But if this conditional is a tautology, then any substitution instance of it is also a tautology. So, for example, $[[(R \vee \sim S) \,\&\, (S \supset. T)] \supset (R \vee \supset S)]$ is a tautology. But then, $[(R \vee \sim S) \,\&\, (S \supset T)]$ tautologically implies $(R \vee \sim S)$. So we have obtained a new tautological implication from I2. What we have done is formed substitution instances of $(P \,\&\, Q)$ and P in which we substitute the same things for the same atomic parts in both formulas. This will be generally true; that is, *if one formula tautologically implies a second, then a substitution instance of the antecedent of this implication tautologically implies any substitution instance of the consequent that results from making the same substitutions as in the antecedent.* A simpler way to express this is the following: given any two formulas, P and Q, if P tautologically implies Q, then if R and S are formulas so constructed that $(R \supset S)$ is a substitution instance of $(P \supset Q)$, then R tautologically implies S. Therefore, we can think of the tautological implications listed in Section II.7 as implication schemas, because any substitution instance of one of those tautological implications is also a tautological implication. From I8, for example, we obtain the implication $\sim((S \supset T) \supset \sim\sim R)$ ▯– $(S \supset T)$, by substituting $(S \supset T)$ for P and $\sim \sim R$ for Q. And from I5 we obtain:

$[P \supset (S \equiv \sim R)]$ ▯– $([P \supset (S \equiv \sim R)] \vee (P \,\&\, Q))$

We must be careful in constructing new implications from old that the same substitutions are made in both the antecedent and consequent of the implication. For example, the implication $[(R \,\&\, S) \supset T]$,

$(R \mathbin{\&} S)$ ▯– Q is not a substitution instance of 19, because we have not substituted the same thing for Q in both the antecedent and the consequent.

Rather than talking about substitution instances, there is a simpler way of reading the implication schemes I1–I14. For example, we can read I1 as "Given any formulas, P and Q, $(P \mathbin{\&} Q)$ implies P." And I8 can be read "Given any formulas, P and Q, $\sim(P \supset Q)$ implies $\sim Q$." This is a simple way of saying that any substitution instances of these implications are also implications.

The tautological equivalences we listed in Section II.7 can also be thought of as equivalence schemes rather than particular equivalences. Any substitution instance of one of those equivalences will also be an equivalence. Or, more precisely, *if two formulas are tautologically equivalent, then any substitution instances of those formulas that result from making the same substitutions in both will also be tautologically equivalent.* Thus, from E2, which says that $\sim(P \vee Q)$ is equivalent to $(\sim P \mathbin{\&} \sim Q)$, we can conclude that $\sim[(P \mathbin{\&} Q) \vee (R \supset S)]$ is equivalent to

$$[\sim(P \mathbin{\&} Q) \mathbin{\&} \sim(R \supset S)]$$

This results from the fact that two formulas are tautologically equivalent if, and only if, each tautologically implies the other; so if we make the same substitutions in both, the resulting substitution instances will each imply the other, making them also tautologically equivalent.

II.8 EXERCISES

A. For each of the following pairs of formulas, state whether the formula after the slash is a substitution instance of the formula before the slash, and, if it is, what substitutions were performed.

1. $(P \vee \sim Q) \,/\, [P \vee \sim(R \supset S)]$
2. $(\sim P \supset Q) \,/\, [\sim(S \mathbin{\&} \sim T) \supset \sim(S \mathbin{\&} \sim T)]$
3. $(P \vee \sim Q) \,/\, [(R \vee S) \vee (\sim R \mathbin{\&} S)]$
4. $[P \supset (Q \equiv P)] \,/\, [\sim\sim S \supset ((P \equiv Q) \equiv \sim\sim S)]$
5. $\{(R \supset \sim S) \supset [(\sim R \vee S) \equiv [(R \mathbin{\&} \sim\sim S) \supset \sim(S \vee R)]]\} \,/$ $\{[(T \mathbin{\&} \sim R) \supset \sim\sim R] \supset \{[\sim(T \mathbin{\&} \sim R) \vee \sim R]$ $\equiv [[(T \mathbin{\&} \sim R) \mathbin{\&} \sim\sim\sim R] \supset \sim[\sim R \vee (T \mathbin{\&} \sim R)]]\}\}$
6. $P \,/\, [Q \supset (P \supset \sim Q)]$
7. $(R \vee \sim R) \,/\, [(P \vee \sim Q) \vee \sim R]$

8. $[P \supset (Q \supset {\sim}R)]\ /\ [{\sim}R \supset (P \supset Q)]$
9. $\{[P \supset (Q \supset R)] \supset [(P \,\&\, Q) \supset R]\}\ /$
   $\{[(P \supset Q) \supset [(Q \supset R) \supset (P \supset R)]]$
   $\supset [[(P \supset Q) \,\&\, (Q \supset R)] \supset (P \supset R)]\}$
10. $[(P \supset Q) \equiv {\sim}(P \,\&\, {\sim}Q)]\ /$
    $\{[P \supset {\sim}(Q \,\&\, {\sim}P)] \equiv {\sim}[P \,\&\, (Q \,\&\, {\sim}P)]\}$

B. State whether each of the following purported tautological implications and tautological equivalences can be obtained by substitution from one of the tautological implications or tautological equivalences listed in Section II.7, and, if so, from which one.

1. $\{(P \,\&\, Q) \supset [R \supset (P \,\&\, (Q \,\&\, R))]\}$
   eq. $[(P \,\&\, (Q \,\&\, R)) \supset (P \,\&\, (Q \,\&\, R))]$
2. $\{[(P \equiv Q) \supset {\sim}(Q \equiv {\sim}P)] \,\&\, [{\sim}(Q \equiv {\sim}P) \supset (P \equiv Q)]\}$
   eq. $[(P \equiv Q) \equiv {\sim}(Q \equiv {\sim}P)]$
3. $[\{(P \supset {\sim}T) \,\&\, [R \vee {\sim}(P \supset Q)]\} \vee \{(Q \equiv {\sim}{\sim}R)$
   $\&\, [R \vee {\sim}(P \supset Q)]\}]$ eq. $\{[(P \supset {\sim}T) \vee (Q \equiv {\sim}{\sim}R)]$
   $\&\, [R \vee {\sim}(P \supset Q)]\}$
4. $[{\sim}(P \supset Q) \vee {\sim}(Q \supset P)]$ eq. $[(P \supset Q) \supset {\sim}(Q \supset P)]$
5. ${\sim}[(P \supset {\sim}Q) \vee (Q \supset {\sim}P)]$ eq. $[{\sim}(P \supset {\sim}Q) \,\&\, {\sim}(Q \supset {\sim}P)]$
6. ${\sim}(Q \vee {\sim}R) \models [(Q \vee {\sim}R) \supset {\sim}(R \,\&\, Q)]$
7. $[(P \supset Q) \,\&\, (Q \,\&\, R)] \models (Q \supset R)$
8. ${\sim}(P \equiv Q),\ [(P \equiv Q) \vee (P \equiv {\sim}Q)] \models (P \equiv {\sim}Q)$
9. ${\sim}[(R \,\&\, {\sim}S) \supset {\sim}R] \models {\sim}{\sim}R$
10. $[(Q \,\&\, {\sim}R) \supset (R \vee P)],\ [(Q \,\&\, {\sim}R) \vee (P \,\&\, {\sim}Q)],$
    $[(P \,\&\, {\sim}Q) \supset (R \vee P)] \models (R \vee P)$

C.* Show that the following statements are true.

1. Any substitution instance of a truth-functionally inconsistent formula must be truth-functionally inconsistent.
2. It is possible for a formula to be truth-functionally consistent but have a substitution instance that is truth-functionally inconsistent.
3. It is possible for a formula not to be a tautology but to have a substitution instance that is a tautology.

II.9 TAUTOLOGICAL REPLACEMENT

In constructing substitution instances, we were only allowed to substitute formulas for the *atomic* parts of the formulas. In replacement, this restriction and the restriction about replacing a formula by the same formula throughout are relaxed. Replacement is simply the process of replacing a part, either atomic or molecular, of one formula by another formula. Thus, $(P \,\&\, Q)$ is the formula that results from replacing $(R \supset (S \,\&\, \sim T))$ by Q in $(P \,\&\, (R \supset (S \,\&\, \sim T)))$. And $(P \vee Q)$ is the formula that results from replacing the first occurrence of $(S \supset \sim T)$ in $[(S \supset \sim T) \vee (S \supset \sim T)]$ by P and the second occurrence by Q. Note that you do not have to replace every occurrence of a part by the same formula.

Generally, when we replace a part of one formula by another formula we get nothing interesting. But if we put certain restrictions on what we replace by what, the situation improves. The replacements with which we are going to be concerned are those in which we replace a part of a formula by another formula that is tautologically equivalent to what it replaces. Thus, for example, using E11 we might replace an occurrence of $(P \,\&\, Q)$ by $(Q \,\&\, P)$ in the formula $[(P \,\&\, Q) \equiv \sim\sim(P \,\&\, Q)]$, obtaining $[(Q \,\&\, P) \equiv \sim\sim(P \,\&\, Q)]$. One formula can be obtained from another formula by *tautological replacement* if the first formula is obtained from the second formula by replacing one of the parts of the second formula by a formula to which it is tautologically equivalent. Then the formula $[(Q \,\&\, P) \equiv \sim\sim(P \,\&\, Q)]$ is obtained by tautological replacement from $[(P \,\&\, Q) \equiv \sim\sim(P \,\&\, Q)]$.

Notice that if one formula can be obtained from a second formula by tautological replacement, then the second formula can also be obtained from the first formula by tautological replacement, just by putting the original part back in for what it was replaced by. For example, we can obtain $[(Q \,\&\, P) \equiv \sim\sim(P \,\&\, Q)]$ by tautological replacement from $[(P \,\&\, Q) \equiv \sim\sim(P \,\&\, Q)]$ by replacing the first $(P \,\&\, Q)$ by $(Q \,\&\, P)$. Then we can get $[(P \,\&\, Q) \equiv \sim\sim(P \,\&\, Q)]$ back again by tautological replacement by simply putting $(P \,\&\, Q)$ back in for $(Q \,\&\, P)$.

What makes tautological replacement interesting is that it yields tautologically equivalent formulas; that is, *if one formula can be obtained from a second by tautological replacement, then they are tautologically equivalent.* This should be clear from thinking about their truth tables. The truth value of a formula is determined by the truth values of its parts. If one formula is obtained from another by tautological replacement, then they are identical except for some parts that are tautologically equivalent. Those parts that are tautologically equivalent will always have the same truth values, and the rest of the formulas will have the same truth values because they are identical, so the formulas themselves must have the same truth values. Consider the formulas

$$[(P \,\&\, Q) \equiv \sim\sim(P \,\&\, Q)]$$

and $[(Q \& P) \equiv \sim\sim(P \& Q)]$. On any given line of their truth tables, $(P \& Q)$ and $(Q \& P)$ will have the same value because they are tautologically equivalent, and $\sim\sim(P \& Q)$ is the same in both formulas so it will have the same truth value in each. Thus each side of the first biconditional will have the same truth value as that same side in the second biconditional; therefore, both biconditionals will have the same truth value.

Recall that any formula that is tautologically equivalent to a tautology must itself be a tautology. Any formula that is obtained from a tautology by a tautological replacement will be equivalent to that tautology. Therefore, any formula, which can be obtained from a tautology by tautological replacement will itself be a tautology. This gives us another way to establish that formulas are tautologies without working out truth tables. If we already know that one formula is a tautology, we can establish that a second formula is also a tautology if we can show that the second formula can be obtained from the first by tautological replacement. Suppose we want to show that $(P \supset \sim\sim P)$ is a tautology. We know that $(P \supset P)$ is a tautology, and we can obtain $(P \supset \sim\sim P)$ from $(P \supset P)$ by tautological replacement of $\sim\sim P$ for the second occurrence of P (using E1). Thus $(P \supset \sim\sim P)$ is a tautology.

We can combine substitution and tautological replacement to establish that many complex formulas are tautologies. If we want to show that $[(R \vee S) \supset ((\sim R \& \sim S) \supset \sim(R \& \supset S))]$ is a tautology, rather than construct a truth table we might proceed as follows: First we observe that Q tautologically implies $(P \supset Q)$ (Rule I7), and so $[Q \supset (P \supset Q)]$ is a tautology. Then by substitution,

$$[(R \vee S) \supset ((R \& \sim S) \supset (R \vee S))]$$

is a tautology. Here we have substituted $(R \vee S)$ for Q and $(R \& \sim S)$ for P. By Rule E16, $[(R \& \sim S) \supset (R \vee S)]$ is tautologically equivalent to $[\sim(R \vee S) \supset \sim(R \& \sim S)]$, so by tautological replacement, $[(R \vee S) \supset [\sim(R \vee S) \supset \sim(R \& \sim S)]]$ is a tautology. By Rule E2, $\sim(R \vee S)$ is tautologically equivalent to $(\sim R \& \sim S)$, so by tautological replacement again,

$$[(R \vee S) \supset [(\sim R \& \sim S) \supset \sim(R \& \sim S)]]$$

is a tautology.

The difference between substitution and tautological replacement is worth reviewing. In substitution, we replace the *atomic* parts of a formula by other formulas to construct a more complicated formula. A substitution instance of a formula will not generally be equivalent to that formula. However, a substitution instance of a tautology will be a tautology. On the other hand, in tautological replacement we replace one part of a formula (*not necessarily an atomic part*) by another formula to which it is equivalent, and in this way we construct a formula that *is* equivalent to the original formula.

II.9 EXERCISES

For each of the following pairs of formulas, tell whether they can be obtained from one another by a single application of tautological replacement using a substitution instance of one of the tautological equivalences E1–E21. If they can, which equivalence and what substitution instance is used?

1. $[P \supset (Q \supset R)] \,/\, [P \supset ({\sim}Q \vee R)]$
2. $[(R \,\&\, {\sim}S) \equiv {\sim}T] \,/\, \{[(R \,\&\, {\sim}S) \supset {\sim}T] \,\&\, [{\sim}T \supset (R \,\&\, {\sim}S)]\}$
3. $[(P \vee {\sim}Q) \equiv {\sim}(R \,\&\, {\sim}S)] \,/\, [(P \vee {\sim}Q) \equiv ({\sim}R \,\&\, {\sim}{\sim}S)]$
4. ${\sim}(P \,\&\, {\sim}Q) \,/\, ({\sim}P \vee Q)$
5. ${\sim}\{[{\sim}Q \,\&\, (R \vee S)] \vee [{\sim}Q \,\&\, (P \supset {\sim}S)]\} \,/$
   ${\sim}\{{\sim}Q \,\&\, [(R \vee S) \vee (P \supset {\sim}S)]\}$
6. $\{[(R \,\&\, {\sim}S) \supset P] \supset {\sim}Q\} \,/\, \{[R \supset ({\sim}S \supset P)] \supset {\sim}Q\}$
7. $[(P \,\&\, Q) \supset ({\sim}Q \,\&\, {\sim}P)] \,/\, [{\sim}(P \,\&\, Q) \supset {\sim}({\sim}Q \,\&\, {\sim}P)]$
8. $\{[R \equiv (S \,\&\, {\sim}S)] \vee {\sim}T\} \,/\, \{[R \equiv {\sim}(S \supset S)] \vee {\sim}T\}$
9. $\{(P \supset {\sim}Q) \,\&\, [({\sim}Q \,\&\, T) \vee {\sim}S]\} \,/$
   $\{[(P \supset {\sim}Q) \,\&\, ({\sim}Q \,\&\, T)] \vee {\sim}S\}$
10. $\{[P \supset (Q \supset R)] \equiv {\sim}[(S \,\&\, {\sim}Q) \vee ({\sim}S \,\&\, Q)]\} \,/$
    $\{[P \supset (Q \supset R)] \equiv {\sim}(S \equiv {\sim}Q)]\}$
11. $\{[P \supset {\sim}(R \equiv {\sim}S)] \,\&\, {\sim}T\} \,/\, \{[P \supset (R \equiv S)] \,\&\, {\sim}T\}$

II.10 DERIVATIONS IN THE PROPOSITIONAL CALCULUS

In principle, we can always tell if a formula is tautologically implied by a set of formulas by constructing a truth table for the corresponding conditional. But, as we have seen, this can become extremely unwieldy if there are more than three atomic parts involved. For example, if there are eight atomic parts among the formulas, then the truth table for the corresponding conditional would be 256 lines long. Thus it is advantageous to have a way of shortcutting this procedure. We have seen how this can be done using rules about substitution and tautological replacement. In this section we will go further and construct a set of rules that will always, by themselves, be sufficient to derive any tautological consequence from a set of formulas that tautologically implies it. With these rules, we will never have to use truth tables; they will enable us to construct sound argument forms, and these sound argument forms are called *derivations in the propositional calculus*, or simply *PC derivations*. In addition to providing an alternative to truth tables, PC derivations will also constitute a useful tool for analyzing many familiar patterns of informal reasoning.

Before saying precisely what a PC-derivation is, let us look at an example of how we might construct a sound argument form in the propositional calculus. Suppose we want to show that $\sim(P \equiv Q)$ is tautologically implied by $(P \supset \sim Q)$ and $(\sim Q \supset P)$. We might construct an argument form as follows:

(1)	1.	$(P \supset \sim Q)$	premise
(2)	2.	$(\sim Q \supset P)$	premise
(1)	3.	$\sim(P \,\&\, \sim\sim Q)$	tautological replacement from 1, using E5.
(1)	4.	$\sim(P \,\&\, Q)$	tautological replacement from 3, using E1.
(2)	5.	$\sim(\sim Q \,\&\, \sim P)$	tautological replacement from 2, using E5.
(2)	6.	$\sim(\sim P \,\&\, \sim Q)$	tautological replacement from 5, using E11.
(1, 2)	7.	$[\sim(P \,\&\, Q) \,\&\, \sim(\sim P \,\&\, \sim Q)]$	tautologically implied by 4 and 6, using I13.
(1, 2)	8.	$\sim[(P \,\&\, Q) \vee (\sim P \,\&\, \sim Q)]$	tautological replacement from 7, using E2.
(1, 2)	9.	$\sim(P \equiv Q)$	tautological replacement from 8, using E8.

The structure of this argument form consists of consecutively numbered lines. We began by taking two premises. Then each succeeding line was obtained from previous lines by either tautological replacement or tautological implication. To the right of each line we indicate how we got it, and from what lines it was obtained. In parentheses on the left side of each line we indicate from which premises that line was obtained. For example, (2) to the left of line 5 indicates that line 5 was obtained from the premise on line 2, and (1, 2) to the left of line 7 indicates that line 7 was obtained from the premises on lines 1 and 2. These numbers are called the *premise numbers* of the line. The formulas written on those lines whose numbers are the premise numbers of a later line are called the *premises* of that later line.

One further thing that needs explanation is why we wrote (1) to the left of line 1 and (2) to the left of line 2. This amounts to saying that each premise is its own premise. We do this because we want to be able to say that each line of a sound argument form in the propositional calculus is tautologically implied by the premises whose numbers are listed to the left of it. And we also want to be able to say that if a line has no premise numbers listed to its left, then the formula on that line is a tautology. This means that we cannot leave the column to the left of the line number blank for a premise, since that would indicate that the premise is a tautology; we must put something in that column. Furthermore, we want the number we write in that column to be the number of a premise that tautologically implies the formula we are writing on that line. Every formula tautologically implies itself, so we can simply take that same line number and write it to the left as a premise number. Then, as the premise tautologically implies itself, it will remain true that every formula is tautologically implied by its premises.

PC derivations will be constructed on the model of the above argument form. That is, a PC derivation will consist of a sequence of lines each of which consists of a set of premise numbers, a line number, a formula, and an explanation of how this line was obtained. We will have a short list of four rules (such as a rule allowing us to make tautological replacements), and each step of a PC derivation must be made in accordance with one of these rules. Any finite sequence of consecutively numbered lines in which each line can be introduced using one of these rules will be a PC derivation. The rules will now be listed, and then discussed individually:

RULE P: PREMISE INTRODUCTION: Any formula can be written on any line of a PC derivation provided that we let the premise number be the line number of that line.

RULE I: IMPLICATION: If, according to one of the implication schemes I1–I15, a set of formulas appears on one or more lines of a PC derivation that tautologically implies a further formula, then we can write that further formula on any later line of the PC derivation, taking as premise numbers all of the premise numbers of the set of formulas.

RULE C: CONDITIONALIZATION: If some formula, Q, appears on a line of a PC derivation, and a formula, P, is among the premises of Q, then on any later line of the PC derivation we can write the conditional $(P \supset Q)$, taking for premise numbers all the premise numbers of Q except for the line number of P.

RULE R: TAUTOLOGICAL REPLACEMENT: If one formula appears on a line of a PC derivation, and a second formula can be obtained from the first by tautological replacement using one of the equivalences E1–E21, then we can write the second formula on any later line of the PC derivation, taking all of the premise numbers of the first formula for the premise numbers.

These rules are called *rules of inference for the propositional calculus*, because they allow us to infer that if one formula or set of formulas is true, then another formula that can be obtained using these rules is also true.[6]

To illustrate the use of these rules in constructing PC derivations, let us look at a simple example of a PC derivation:

(1)	1.	P	P
(2)	2.	$\sim Q$	P
(1, 2)	3.	$(P \,\&\, \sim Q)$	I, 1, 2 (I13)
(1, 2)	4.	$\sim(P \supset Q)$	R, 3 (E5)
(1)	5.	$[\sim Q \supset \sim(P \supset Q)]$	C, 2, 4
	6.	$\{P \supset [\sim Q \supset \sim(P \supset Q)]\}$	C, 1, 5

On lines 1 and 2 we used Rule P, taking the premises P and $\sim Q$, respectively. On line 3 we conjoined the premises, using Rule I and the implication scheme I13. On line 4 we used Rule R to replace $(P \,\&\, \sim Q)$ by $\sim(P \supset Q)$, to which it is tautologically equivalent by E5. On line 5 we used Rule C to form the conditional

$$[\sim Q \supset \sim(P \supset Q)]$$

thus eliminating the premise number 2. On line 6 we used Rule C again, eliminating the premise number 1. Note that line 6 has no premise numbers, which means it is a tautology.

[6] For the time being, the student should simply take it on faith that whenever a formula can be obtained from a set of premises using these rules of inference, then that formula is tautologically implied by those premises. That this is so will be established in Section II.11.

A PC derivation whose last line is a formula without any premise numbers is said to be a PC derivation *of that formula*; for example, the above PC derivation is a PC derivation of $[P \supset (\sim Q \supset \sim(P \supset Q))]$. If there is a PC derivation of a formula, that formula is said to be a *theorem of the propositional calculus*; thus $[P \supset (\sim Q \supset \sim(P \supset Q))]$ is a theorem of the propositional calculus. A PC derivation whose last line is a formula with premise numbers is said to be a PC derivation *of* that formula *from* those premises. Thus, if we had stopped the above derivation on line 5, we would have had a PC derivation of $[\sim Q \supset \sim(P \supset Q)]$ from P. It will be shown after a while that a formula of the propositional calculus is a tautology if, and only if, it is a theorem of the propositional calculus; and a formula is tautologically implied by a set of formulas if, and only if, there is a PC derivation of it from those premises.

Now, let us look at the rules of inference. Rule P needs little discussion, except to note that you can take *anything at all* as a premise. The only thing that restricts what one takes as a premise is just what one is trying to derive. In any given case, most premises will not be of any use, so it will be usefulness that dictates what premises one should take. But anything at all *can* be introduced as a premise. However, the point in constructing a PC derivation of a formula from certain premises is to show that the formula is tautologically implied by those premises. Thus it is important that the derivation constructed has as its last line a line on which the desired conclusion is written and whose *only* premise numbers are the numbers of the premises from which the conclusion was to be obtained. If there are any additional premise numbers, then we have not shown that the conclusion is tautologically implied by the given premises, and so have not done what we set out to do.

It must be emphasized that Rule I can only be applied to an *entire line*. If, for example, $(P \,\&\, Q)$ is the formula on one line of a PC derivation, then we can use Rule I with I2 to obtain Q on a later line. But if $[(P \,\&\, Q) \supset R]$ appears on a line of a PC derivation, we cannot use Rule I with I2 to obtain anything from it. In particular, we cannot obtain $(Q \supset R)$ by Rule I, using I2. In fact, $(Q \supset R)$ is simply not implied by $[(P \,\&\, Q) \supset R]$.

Rule R is different from Rule I in that it need not be applied to an entire line. Using Rule R we can transform *any part* of a formula appearing on a line into something to which it is tautologically equivalent. For instance, if $[R \supset (P \,\&\, Q)]$ appears on a line, we might use Rule R with E11 to transform it into $[R \supset (Q \,\&\, P)]$. Of course, Rule R can also be used to transform the entire line into something else: if the entire line were $(P \,\&\, Q)$, we could again use Rule R with E11 to obtain $(Q \,\&\, P)$ as a new line. It is very important to keep this difference straight. Rule I can only be used on an entire line, while Rule R can be used on any part of a line.

Rule C is essentially just a rule that allows us to say that if we have derived a formula from a premise, then if that premise is true the formula is true. The following is a simple illustration of the use of Rule C:

(1)	1.	P	P
(1)	2.	$(P \vee Q)$	I, 1 (I3)
	3.	$[P \supset (P \vee Q)]$	C, 1, 2

From the premise P we derived the conclusion $(P \vee Q)$. Thus, if P is true, then $(P \vee Q)$ is true. But this means that $[P \supset (P \vee Q)]$ is true; and this is just what we get on line 3 using Rule C. Rule C allows us to eliminate a premise by writing it explicitly as the antecedent of a conditional, thus saying that *if* that premise is true, *then* the conclusion we derived from it is true.

Another example of the use of Rule C is the following:

(1)	1.	P	P
(2)	2.	Q	P
(1, 2)	3.	$(P \,\&\, Q)$	I, 1, 2 (I13)
(1)	4.	$[Q \supset (P \,\&\, Q)]$	C, 2, 3
	5.	$\{P \supset [Q \supset (P \,\&\, Q)]\}$	C, 1, 4

From the premises P and Q we derived the conclusion $(P \,\&\, Q)$. So $(P \,\&\, Q)$ follows from premises (1) and (2). Then, eliminating premise (2), we can say that—given premise (1)—*if* premise (2) is true then $(P \,\&\, Q)$ is true; that is, $[Q \supset (P \,\&\, Q)]$ is true. So we can infer $[Q \supset (P \,\&\, Q)]$ from the premise P. Then using Rule C again, we can say that *if* P is true, then $[Q \supset (P \,\&\, Q)]$ is true; that is,

$$\{P \supset [Q \supset (P \,\&\, Q)]\}$$

is true. This latter formula now no longer depends upon any premises, because we have taken the only premise of the consequent and written it explicitly as the antecedent of the conditional. The process of using Rule C to eliminate a premise and write it as the antecedent of a conditional is called *conditionalization.* Thus, on line 4 we say that we are conditionalizing $(P \& Q)$ *with respect to* Q.

One thing that must be emphasized here about Rule C is that *you can only conditionalize a formula with respect to one of its premises.* If a formula does not occur among the premises of a given line, then you cannot conditionalize that line with respect to that formula. For instance, the use of Rule C on line 5 of the following purported derivation is incorrect:

(1)	1.	$(P \& Q)$	P
(1)	2.	Q	I, 1 (I2)
(1)	3.	$(Q \vee P)$	I, 2 (I3)
(1)	4.	P	I, 1 (I1)
	5.	$[(Q \vee P) \supset P]$	C, 3, 4

We cannot conditionalize in this way, because $(Q \vee P)$ is not one of the premises of line 4.

Now let us look at some more examples of PC derivations. The student should work through each of these carefully and be sure that he knows how each line was obtained. However, he should not worry about *why* each line was introduced when it was. A discussion of strategies involved in constructing PC derivations will follow, but for the time being the student is only asked to be able to verify that the rules are being used correctly.

To derive P from $({\sim}P \supset P)$:

(1)	1.	$({\sim}P \supset P)$	P
(1)	2.	$({\sim}{\sim}P \vee P)$	R, 1 (E4)
(1)	3.	$(P \vee P)$	R, 2 (E1)
(1)	4.	P	R, 3 (E18)

To derive $({\sim}P \equiv {\sim}Q)$ from $(P \equiv Q)$:

(1)	1.	$(P \equiv Q)$	P
(2)	2.	${\sim}P$	P
(1)	3.	$[(P \supset Q) \& (Q \supset P)]$	R, 1 (E7)
(1)	4.	$(Q \supset P)$	I, 3 (I2)
(1, 2)	5.	${\sim}Q$	I, 2, 4 (I10)
(1)	6.	$({\sim}P \supset {\sim}Q)$	C, 2, 5
(7)	7.	${\sim}Q$	P
(1)	8.	$(P \supset Q)$	I, 3 (I1)
(1, 7)	9.	${\sim}P$	I, 7, 8 (I10)

(1)	10.	$(\sim Q \supset \sim P)$	C, 7, 9
(1)	11.	$[(\sim P \supset \sim Q) \& (\sim Q \supset \sim P)]$	I, 6, 10 (I13)
(1)	12.	$(\sim P \equiv \sim Q)$	R, 11 (E7)

To derive $(P \equiv R)$ from $(P \equiv Q)$ and $(Q \equiv R)$:

(1)	1.	$(P \equiv Q)$	P
(2)	2.	$(Q \equiv R)$	P
(1)	3.	$[(P \supset Q) \& (Q \supset P)]$	R, 1 (E7)
(2)	4.	$[(Q \supset R) \& (R \supset Q)]$	R, 2 (E7)
(1)	5.	$(P \supset Q)$	I, 3 (I1)
(2)	6.	$(Q \supset R)$	I, 4 (I1)
(1, 2)	7.	$(P \supset R)$	I, 5, 6 (I12)
(2)	8.	$(R \supset Q)$	I, 4 (I2)
(1)	9.	$(Q \supset P)$	I, 3 (I2)
(1, 2)	10.	$(R \supset P)$	I, 8, 9 (I12)
(1, 2)	11.	$[(P \supset R) \& (R \supset P)]$	I, 7, 10 (I13)
(1, 2)	12.	$(P \equiv R)$	R, 11 (E7)

To derive $[\sim(P \equiv Q) \equiv Q]$ from $\sim P$:

(1)	1.	$\sim P$	P
(2)	2.	Q	P
(3)	3.	$(P \& Q)$	P
(3)	4.	P	I, 3 (I1)
	5.	$[(P \& Q) \supset P]$	C, 3, 4
(1)	6.	$\sim(P \& Q)$	I, 1, 5 (I10)
(7)	7.	$(\sim P \& \sim Q)$	P
(7)	8.	$\sim Q$	I, 7 (I2)
	9.	$[(\sim P \& \sim Q) \supset \sim Q]$	C, 7, 8
(2)	10.	$\sim\sim Q$	R, 2 (E1)
(2)	11.	$\sim(\sim P \& \sim Q)$	I, 9, 10 (I10)
(1, 2)	12.	$[\sim(P \& Q) \& \sim(\sim P \& \sim Q)]$	I, 6, 11 (I13)
(1, 2)	13.	$\sim[(P \& Q) \vee (\sim P \& \sim Q)]$	R, 12 (E2)
(1, 2)	14.	$\sim(P \equiv Q)$	R, 13 (E8)
(1)	15.	$[Q \supset \sim(P \equiv Q)]$	C, 2, 14
(16)	16.	$\sim(P \equiv Q)$	P
(16)	17.	$\sim[(P \& Q) \vee (\sim P \& \sim Q)]$	R, 16 (E8)
(16)	18.	$[\sim(P \& Q) \& \sim(\sim P \& \sim Q)]$	R, 17 (E2)
(16)	19.	$\sim(\sim P \& \sim Q)$	I, 18 (I2)
(20)	20.	$\sim Q$	P
(1, 20)	21.	$(\sim P \& \sim Q)$	I, 1, 20 (I13)
(1)	22.	$[\sim Q \supset (\sim P \& \sim Q)]$	C, 20, 21
(1, 16)	23.	$\sim\sim Q$	I, 19, 22 (I10)
(1, 16)	24.	Q	R, 23 (E1)

(1)	25.	$[\sim(P \equiv Q) \supset Q]$	C, 16, 24
(1)	26.	$\{[\sim(P \equiv Q) \supset Q]$ $\& [Q \supset \sim(P \equiv Q)]\}$	I, 15, 25 (I13)
(1)	27.	$[\sim(P \equiv Q) \equiv Q]$	R, 26 (E7)

PC derivations can be very handy tools for establishing the soundness of arguments expressed in English; for example, let us look at Argument E in Chapter I, which went as follows:

1. If the river floods, then our entire wheat crop will be destroyed.
2. The river will flood if, and only if, there is an early thaw in the mountains.
3. But in any case there will be heavy rains later in the summer.

---

4. Therefore, if there is an early thaw in the mountains, then our entire wheat crop will be destroyed, and the community will be bankrupt unless there are heavy rains later in the summer.

We can symbolize the form of this argument in the propositional calculus as follows:

1. $(R \supset W)$
2. $(R \equiv T)$
3. S

---

4. $(T \supset [W \& (\sim B \supset S)])$

We could evaluate this argument by constructing its corresponding conditional and then writing out a truth table, but such a truth table would be 32 lines long. It is much easier to construct a PC derivation of the conclusion from the premises as follows:

(1)	1.	$(R \supset W)$	P
(2)	2.	$(R \equiv T)$	P
(3)	3.	S	P
(4)	4.	T	P
(2)	5.	$[(R \supset T) \& (T \supset R)]$	R, 2 (E7)
(2)	6.	$(T \supset R)$	I, 5 (I2)
(2, 4)	7.	R	I, 4, 6 (I9)
(1, 2, 4)	8.	W	I, 1, 7 (I9)
(3)	9.	$(\sim B \supset S)$	I, 3 (I6)
(1, 2, 3, 4)	10.	$[W \& (\sim B \supset S)]$	I, 8, 9 (I13)
(1, 2, 3)	11.	$(T \supset [W \& (\sim B \supset S)])$	C, 4, 10

"This is all very well," you may reply, "but how does one find derivations?" The answer is that it requires ingenuity. The construction of derivations is not mechanical like the construction of truth tables. So one cannot hope to have a precise set of rules to follow for finding derivations. However, there are definite strategies that are often helpful in constructing PC derivations. We can list different strategies for each type of formula (for example, conjunction, disjunction, and so on).

A Conditionals

We might be confronted with two situations in relation to conditionals. We might be trying to get a conditional as a consequence of something else, or we might be trying to get something else as a consequence of a conditional. These cases are treated separately.

Let us suppose first that we are trying to derive a conditional from something else. The last derivation above illustrates what is generally the simplest way to go about this. In deriving the conditional $(T \supset [W \& (\sim B \supset S)])$ from the premises, what we did was take the antecedent T as a new premise, and then with its help we derived the consequent $[W \& (\sim B \supset S)]$. Then we simply conditionalize to get the conditional. This is almost always the best way to derive a conditional from its premises. Let us look at another example of this same technique. Suppose we want to show that

$$(P \supset [Q \supset (R \supset [P \& (Q \& R)])])$$

is a tautology. Then we can construct a derivation as follows:

(1)	1.	P	P
(2)	2.	Q	P
(3)	3.	R	P
(1, 2, 3)	4.	$[P \& (Q \& R)]$	I, 1, 2, 3 (I15)
(1, 2)	5.	$(R \supset [P \& (Q \& R)])$	C, 3, 4
(1)	6.	$[Q \supset (R \supset [P \& (Q \& R)])]$	C, 2, 5
	7.	$(P \supset [Q \supset (R \supset [P \& (Q \& R)])])$	C, 1, 6

In constructing this derivation, we first took the antecedent of the conditional, P, as a premise, and then sought to derive the consequent $[Q \supset (R \supset [P \& (Q \& R)])]$ from it. But this consequent is also a conditional, so we took its antecedent, Q, as a new premise and then sought to derive its consequent, $(R \supset [P \& (Q \& R)])$. But that consequent is also a conditional, so we took its antecedent, R, as a third premise and then sought to derive its consequent, $[P \& (Q \& R)]$. This we could do immediately using I15. Then we conditionalized in reverse order until we had attached all of the antecedents in their proper places.

Next let us try to derive something from a conditional. There are four common situations of this sort. First, we might have a conditional of the form $(P \supset Q)$, and have some other conditional of the form $(Q \supset R)$. From these, using I12, we can derive $(P \supset R)$. This inference is called *hypothetical syllogism.* If we wished we could introduce a new rule for this, but it would just be a special case of Rule I. An example of the use of hypothetical syllogism is found in the following derivation in which $(\sim P \vee R)$ is derived from the premises $(\sim P \vee Q)$ and $(\sim Q \vee R)$:

(1)	1.	$(\sim P \vee Q)$	P
(2)	2.	$(\sim Q \vee R)$	P
(1)	3.	$(P \supset Q)$	R, 1 (E4)
(2)	4.	$(Q \supset R)$	R, 2 (E4)
(1, 2)	5.	$(P \supset R)$	I, 3, 4 (I12) (hypothetical syllogism)
(1, 2)	6.	$(\sim P \vee R)$	R, 5, (E4)

A second common case is where we have a conditional $(P \supset Q)$ on one line and on another line, the antecedent P. Then by I9 we can derive Q. This inference is called *modus ponens*. Notice that this is what we used in getting from lines 4 and 6 to line 7 in the derivation on page 56. A second example would be the following:

(1)	1.	$(P \& Q)$	P
(2)	2.	$[Q \supset (R \& \sim S)]$	P
(1)	3.	Q	I, 1 (I2)
(1, 2)	4.	$(R \& \sim S)$	I, 2, 3 (I9) (*modus ponens*)
(1, 2)	5.	$\sim S$	I, 4 (I2)

A third common case is that in which we have a conditional $(P \supset Q)$ on one line, and then on another line, $\sim Q$, the negation of its consequent. Then using I10 we can derive $\sim P$, the negation of its antecedent. This inference is called *modus tollens*. The following contains an example of *modus tollens*:

(1)	1.	$[Q \supset (R \& \sim S)]$	P
(2)	2.	S	P
(2)	3.	$(\sim R \vee S)$	I, 2 (I4)

(2)	4.	$(\sim R \vee \sim\sim S)$	R, 3 (E1)
(2)	5.	$\sim(R \,\&\, \sim S)$	R, 4 (E2)
(1, 2)	6.	$\sim Q$	I, 1, 5 (I10) (*modus tollens*)

A fourth common case involves an inference of one of the following forms:

$$(P \supset Q) \dashv\vdash (\sim Q \supset \sim P)$$
$$(P \supset \sim Q) \dashv\vdash (Q \supset \sim P)$$
$$(\sim P \supset Q) \dashv\vdash (\sim Q \supset P)$$
$$(\sim P \supset \sim Q) \dashv\vdash (Q \supset P)$$

These four inferences are all called *contraposition*. We can do either the first or the fourth immediately using E16 with Rule R. The second and the third require a few intermediate steps. Consider the second; from $(P \supset \sim Q)$ we can infer $(\sim\sim P \supset \sim Q)$ by Rule R, using E1. Then by the fourth form of contraposition, we obtain $(Q \supset \sim P)$. The third form is similar.

B Conjunctions

First consider the case in which we are trying to derive a conjunction from something else. The strategy here is quite simple. We simply derive each conjunct separately, and then conjoin them using Rule I and I13. An example of this is found on line 10 of the derivation on page 56.

The case in which we are trying to derive a consequent from a conjunction is equally simple. We use Rule I together with I1 and I2 to get the conjuncts separately, and then we work with them. For example, suppose we want to derive $(\sim R \supset \sim P)$ from

$$[(P \supset Q) \,\&\, (Q \supset R)]$$

We can do this as follows:

(1)	1.	$[(P \supset Q) \,\&\, (Q \supset R)]$	P
(1)	2.	$(P \supset Q)$	I, 1 (I1)
(1)	3.	$(Q \supset R)$	I, 1 (I2)
(1)	4.	$(P \supset R)$	I, 2, 3 (I12) (hypothetical syllogism)
(1)	5.	$(\sim R \supset \sim P)$	R, 4 (E16) (contraposition)

C Disjunctions

Let us first try to derive a disjunction from something else. The simplest case is that in which we can derive one of the disjuncts separately, and then get the disjunction by Rule I, using I3 or I4. But this cannot generally be done. A more common strategy results from recognizing that, by E4, a disjunction is equivalent to a conditional. Suppose we have a disjunction $(P \vee Q)$. By Rule R and E1, it is equivalent to $(\sim\sim P \vee Q)$. Then by Rule R and E4, it is equivalent to $(\sim P \supset Q)$. So we can treat the disjunction just like we did a conditional; that is, take $\sim P$ as a new premise and try to derive Q. Then conditionalize, getting $(\sim P \supset Q)$, and then reverse the above transformations using rule R to obtain $(P \vee Q)$. Suppose we want to derive $(P \vee Q)$ from $[(P \,\&\, \sim Q) \vee (\sim P \,\&\, Q)]$. This might be done as follows:

(1)	1.	$[(P \,\&\, \sim Q) \vee (\sim P \,\&\, Q)]$	P
(2)	2.	$\sim P$	P
(2)	3.	$(\sim P \vee \sim\sim Q)$	I, 2 (I3)
(2)	4.	$\sim(P \,\&\, \sim Q)$	R, 3 (E3)
(1, 2)	5.	$(\sim P \,\&\, Q)$	I, 1, 4 (I11)
(1, 2)	6.	Q	I, 5 (I2)
(1)	7.	$(\sim P \supset Q)$	C, 2, 6
(1)	8.	$(\sim\sim P \vee Q)$	R, 7 (E4)
(1)	9.	$(P \vee Q)$	R, 8 (E1)

On the other hand, if we wish to derive a conclusion from a disjunction, we generally make use of I14. That is, if we want to derive some conclusion R from a disjunction of the form $(P \vee Q)$, we establish the two conditionals $(P \supset R)$ and $(Q \supset R)$ (we derive R from each disjunct), and then use Rule I with I14 to get R. An example of this would be the following:

(1)	1.	$[(P \,\&\, Q) \vee (P \,\&\, R)]$	P
(2)	2.	$(P \,\&\, Q)$	P
(2)	3.	P	I, 2 (I1)
	4.	$[(P \,\&\, Q) \supset P]$	C, 2, 3
(5)	5.	$(P \,\&\, R)$	P
(5)	6.	P	I, 5 (I1)
	7.	$[(P \,\&\, R) \supset P]$	C, 5, 6
(1)	8.	P	I, 1, 4, 7 (I14)

Another important strategy in deriving a conclusion from a disjunction involves the use of I11, disjunctive syllogism; that is, if we can somehow obtain the negation of one of the disjuncts of the disjunction, then we can obtain the other disjunct by I11. (Notice however that if we have the negation of the second disjunct of the disjunction, we must use E12 to invert the order of the disjuncts before we can use I11.) This use of disjunctive syllogism is found in the top derivation on page 60. The premise on line 1 is a disjunction; in order to obtain the conclusion we derived the negation of the first disjunct, on line 4, and then obtained the second disjunct on line 5 by disjunctive syllogism.

D *Negations*

If we want to derive a conclusion from a negation, often the best strategy is to use one of the equivalences E2, E3, E5, or E6 to turn the premise into something else, and then proceed as above. If, for example, the premise is the negation of a conjunction, by E3 we can turn it into a disjunction and then proceed as above for disjunctions. Suppose we want to derive $(\sim P \vee \sim Q)$ from $\sim(P \vee Q)$, we might proceed as follows:

(1)	1.	$\sim(P \vee Q)$	P
(1)	2.	$(\sim P \,\&\, \sim Q)$	R, 1 (E2)
(1)	3.	$\sim P$	I, 2 (I1)
(1)	4.	$(\sim P \vee \sim Q)$	I, 3 (I3)

However, if we are trying to derive a negation from something else, the situation is more complicated. There is no one strategy that works best most of the time. Often we will simply reverse the above procedure, derive a formula to which the negation is tautologically equivalent, and then use Rule R to obtain the negation. There are, however, two other kinds of proofs that deserve special mention. They are both called *reductio ad absurdum*, which means "reduction to absurdity." In a *reductio ad absurdum* proof, we take the negation of what we are trying to derive as a new premise (that is, we suppose that what we are trying to prove is not true) and show that this leads to an absurdity in one of two ways. The first way is as follows: If we are trying to derive P, we take $\sim P$ as a premise (or if we are trying to derive $\sim P$, we might take P as a premise rather than $\sim\sim P$). Then from this we might try to derive P. Having done so, we can conditionalize, obtaining

$(\sim P \supset P)$. By E4, this is equivalent to $(\sim \sim P \vee P)$, and then by Rule R, using E1, we obtain $(P \vee P)$. By E18, this is equivalent to P; so we have derived P. This kind of proof can be schematized as follows:

			⋮	First part of derivation
...	(k)	...		
$k + 1$	$(k + 1)$	$\sim P$	P	Derivation of P from $\sim P$ and other premises
			⋮	
... $k + 1$	(m)	P		
...	$(m + 1)$	$(\sim P \supset P)$	C, $k + 1$, m	
...	$(m + 2)$	$(\sim \sim P \vee P)$	R, $m + 1$ (E4)	
...	$(m + 3)$	$(P \vee P)$	R, $m + 2$ (E1)	
...	$(m + 4)$	P	R, $m + 3$ (E18)	

Here we suppose we already have the first k lines of a derivation, and now we want to obtain P. On the next line, line $(k + 1)$, we introduce $\sim P$ as a new premise. Then on a later line (m), we derive P from the premise $\sim P$ together with some other premises (whose numbers are indicated by the sequence of dots). We then conditionalize and use Rule R to get P following from the other premises alone. Note that we cannot stop on line (m) because, although we have derived P on that line, one of the premises from which it has been derived is $\sim P$. We want to eliminate that premise, so we conditionalize.

The other form of *reductio ad absurdum* proof goes as follows:

			⋮	First part of derivation
...	(k)	...		
$k + 1$	$(k + 1)$	$\sim P$	P	Derivation of $(Q \,\&\, \sim Q)$ from $\sim P$
			⋮	
... $k + 1$	(m)	$(Q \,\&\, \sim Q)$		
...	$(m + 1)$	$[\sim P \supset (Q \,\&\, \sim Q)]$	C, $k + 1$, m	
...	$(m + 2)$	$[\sim \sim P \vee (Q \,\&\, \sim Q)]$	R, $m + 1$ (E4)	
...	$(m + 3)$	$[P \vee (Q \,\&\, \sim Q)]$	R, $m + 2$ (E1)	
...	$(m + 4)$	P	R, $m + 3$ (E17)	

This form of *reductio ad absurdum* starts out like the other form, but rather than deriving P from $\sim P$, we derive something of the form $(Q \,\&\, \sim Q)$ from $\sim P$.

Reductio ad absurdum proofs can be used to derive formulas of any form, not just negations. But they are often useful in deriving negations, so they have been mentioned here. Now let us look at an example of each form:

Suppose first that we want to derive $[(P \supset Q) \supset (P \supset R)]$ from the premise $[P \supset (Q \supset R)]$. We might use the following derivation, which contains within it (steps 3–12) a *reductio ad absurdum* proof of the first kind:

(1)	1.	$[P \supset (Q \supset R)]$	P
(2)	2.	$(P \supset Q)$	P
(3)	3.	$\sim(P \supset R)$	P
(3)	4.	P	I, 3 (I7)
(1, 3)	5.	$(Q \supset R)$	I, 1, 4 (I9) (*modus ponens*)
(2, 3)	6.	Q	I, 2, 4 (I9) (*modus ponens*)
(1, 2, 3)	7.	R	I, 5, 6 (I9) (*modus ponens*)
(1, 2, 3)	8.	$(P \supset R)$	I, 7 (I6)
(1, 2)	9.	$[\sim(P \supset R) \supset (P \supset R)]$	C, 3, 8
(1, 2)	10.	$[\sim\sim(P \supset R) \vee (P \supset R)]$	R, 9 (E4)
(1, 2)	11.	$[(P \supset R) \vee (P \supset R)]$	R, 10 (E1)
(1, 2)	12.	$(P \supset R)$	R, 11 (E18)
(1)	13.	$[(P \supset Q) \supset (P \supset R)]$	C, 2, 12

In this derivation, we began by using the standard technique in deriving conditionals, we took the antecedent, $(P \supset Q)$, as a new premise. From that point through line 12 we used a typical *reductio ad absurdum* proof, then on the last line we conditionalized to get our conditional.

For an example of the second form of *reductio ad absurdum* proof, suppose we want to derive $\sim(Q \supset R)$ from the premises $(P \supset Q)$ and $\sim(P \supset R)$. We might proceed as follows:

(1)	1.	$(P \supset Q)$	P
(2)	2.	$\sim(P \supset R)$	P
(3)	3.	$(Q \supset R)$	P
(1, 3)	4.	$(P \supset R)$	1, 1, 3 (I12) (hypothetical syllogism)
(1, 2, 3)	5.	$[(P \supset R) \& \sim(P \supset R)]$	I, 2, 4 (I13)
(1, 2)	6.	$((Q \supset R) \supset [(P \supset R) \& \sim(P \supset R)])$	C, 3, 5
(1, 2)	7.	$(\sim(Q \supset R) \vee [(P \supset R) \& \sim(P \supset R)])$	R, 6 (E4)
(1, 2)	8.	$\sim(Q \supset R)$	R, 7 (E17)

Steps 3–8 constitute a *reductio ad absurdum* proof of the second kind.

E Biconditionals

Our treatment of biconditionals is very simple. By E7 the biconditional $(P \equiv Q)$ is equivalent to the conjunction

$$[(P \supset Q) \mathbin{\&} (Q \supset P)],$$

so if we wish to derive something from a biconditional, we can first transform the biconditional into a conjunction, using Rule R, and then proceed as with conjunctions. If we wish to derive a biconditional from something else, we can first derive the conjunction to which it is equivalent, and then use Rule R to transform it into the biconditional. An example of this can be found in the derivation on page 55. Or alternatively, we may use E8 to transform the biconditional into a disjunction, and then work with the disjunction.

This completes the list of strategies for the different kinds of formulas. These strategies are not infallible, they are just guides to help you construct derivations, which often requires considerable ingenuity. Unlike truth tables, the construction of derivations is not purely mechanical.

Now let us look at some examples of the combined use of these strategies to construct derivations. Suppose first that we want to derive the conclusion $\sim P$ from the premises $[(P \mathbin{\&} Q) \supset R]$, $(R \supset S)$, $(Q \mathbin{\&} \sim S)$. We might proceed as follows:

(1)	1.	$[(P \mathbin{\&} Q) \supset R]$	P
(2)	2.	$(R \supset S)$	P
(3)	3.	$(Q \mathbin{\&} \sim S)$	P
(3)	4.	Q	I, 3 (I1)
(3)	5.	$\sim S$	I, 3 (I2)
(2, 3)	6.	$\sim R$	I, 2, 5 (I10)
(1, 2, 3)	7.	$\sim (P \mathbin{\&} Q)$	I, 1, 6 (I10)
(1, 2, 3)	8.	$(\sim P \vee \sim Q)$	R, 7 (E3)
(1, 2, 3)	9.	$(\sim Q \vee \sim P)$	R, 8 (E12)
(3)	10.	$\sim \sim Q$	R, 4 (E1)
(1, 2, 3)	11.	$\sim P$	I, 9, 10 (I11)

Now, let us see how this derivation was constructed. We began with our premises. Our third premise was a conjunction, so following the strategy for conjunctions, we obtained each conjunct separately, on lines 4 and 5. Then we noticed that $\sim S$, on line 5, is the negation of the consequent of the conditional on line 2. So by *modus tollens,* we obtained the negation of its antecedent, which is $\sim R$. But $\sim R$ is the negation of the consequent of the conditional on line 1, so again by *modus tollens*, we obtain the negation of its antecedent, which is $\sim(P \,\&\, Q)$. This is a negation, so we follow the strategy for negations, which is to transform them into something else. In this case, we use one of De Morgan's laws, obtaining $(\sim P \vee \sim Q)$. Now, let us stop and consider what conclusion we are attempting to obtain. We want to derive $\sim P$; that is the first disjunct of this disjunction. Thus, if we can obtain $\sim\sim Q$, the negation of the other disjunct, then by disjunctive syllogism we can obtain $\sim P$. But we already have Q on line 4, so we can obtain $\sim\sim Q$ from it by double negation. Then, in order to use I11, which is disjunctive syllogism, we must invert the order of the disjuncts in $(\sim P \vee \sim Q)$ to obtain $(\sim Q \vee \sim P)$, which we do using E12. This completes our derivation.

Let us consider another example. Suppose we want to derive $(S \supset \sim T)$ from the premise

$$\{\sim[P \supset (S \,\&\, T)] \vee [(S \supset \sim P) \,\&\, (T \supset P)]\}$$

We can do this with the following derivation:

(1)	1.	$\{\sim[P \supset (S \,\&\, T)]$ $\vee [(S \supset \sim P) \,\&\, (T \supset P)]\}$	P
(2)	2.	$\sim[P \supset (S \,\&\, T)]$	P
(2)	3.	$[P \,\&\, \sim(S \,\&\, T)]$	R, 2 (E)
(2)	4.	$\sim(S \,\&\, T)$	I, 3 (I2)
(2)	5.	$\sim(S \,\&\, \sim\sim T)$	R, 4 (E1)
(2)	6.	$\sim\sim(S \supset \sim T)$	R, 5 (E5)
(2)	7.	$(S \supset \sim T)$	R, 6 (E1)
	8.	$\{\sim[P \supset (S \,\&\, T)] \supset (S \supset \sim T)\}$	C, 2, 7
(9)	9.	$[(S \supset \sim P) \,\&\, (T \supset P)]$	P
(9)	10.	$(S \supset \sim P)$	I, 9 (I1)
(9)	11.	$(T \supset P)$	I, 9 (I2)
(9)	12.	$(\sim P \supset \sim T)$	R, 11 (E16)
(9)	13.	$(S \supset \sim T)$	I, 10, 12 (I12)
	14.	$\{[(S \supset \sim P) \,\&\, (T \supset P)] \supset (S \supset \sim T)\}$	C, 9, 13
(1)	15.	$(S \supset \sim T)$	I, 1, 8, 14 (I14)

Here we have followed the strategy for deriving a conclusion from a disjunction. The premise is a disjunction, so we tried to derive the conclusion from each disjunct separately, which we did on lines 8 and 14, and then used I14 to obtain the conclusion from our initial premise. We are trying to obtain the conditionals on lines 8 and 14, so we use the strategy for deriving conditionals, which is to take the antecedent as a new premise, and then try to derive the consequent from it. Thus on line 2 we took the first disjunct as a new premise. This disjunct is a negation, so we used our strategy for negations, which is to turn them into something else. We did this using E5. The result is a conjunction. From a conjunction we can obtain each conjunct separately. In this case we only need the second conjunct, which we obtain by I2—$\sim(S \,\&\, T)$. We are looking for $(S \supset \sim T)$. By E5 and E1 this is equivalent to $\sim(S \,\&\, \sim\sim T)$. But this can be obtained from $\sim(S \,\&\, T)$ by double negation. Thus we can derive $(S \supset \sim T)$ from the first disjunct. Next we turn to the second disjunct, taking it as a new premise. It is a conjunction, so we obtain each conjunct separately, using I1 and I2. The thing to observe next is that if we use contraposition on $(T \supset P)$ to obtain $(\sim P \supset \sim T)$, then we will have $(S \supset \sim P)$ and $(\sim P \supset \sim T)$, which by hypothetical syllogism yields our conclusion, $(S \supset \sim T)$.

We may occasionally encounter a case in which we are told to derive a certain conclusion from a given premise *if* it is possible to do so. If it is not possible, how can we show that this is the case without wasting time looking for a derivation where none can be given? For instance, suppose we want to show that $(\sim P \supset \sim Q)$ cannot be derived from $(P \supset Q)$. $(\sim P \supset \sim Q)$ can be derived from $(P \supset Q)$ if, and only if, it is tautologically implied by it; that is, if, and only if, the corresponding conditional $[(P \supset Q) \supset (\sim P \supset \sim Q)]$ is a tautology. So we must establish that the corresponding conditional is not a tautology. The only way we have of doing this is by using a truth table, *but* we do not have to construct an entire truth table. If we can construct a single line of a truth table on which this conditional is false, that will suffice to show that it is not a tautology. To construct such a line of a truth table is just to find truth values that can be assigned to P and Q that will make the conditional false. But for this assignment of truth values to make the corresponding conditional false is just for it to make the premises of the original argument true and the conclusion false. Therefore, the task of showing that a conclusion cannot be derived from certain premises reduces it to finding truth values for the atomic parts of the formulas, which will make all of the premises true and the conclusion false. Thus, in showing that $(\sim P \supset \sim Q)$ cannot be derived from $(P \supset Q)$, what we must do is find truth values for P and Q that make the premise, $(P \supset Q)$, true, and the conclusion, $(\sim P \supset \sim Q)$, false. In order for $(\sim P \supset \sim Q)$ to be false, P must be false and Q true. But this assignment of truth values will also make $(P \supset Q)$ true. So this assignment of truth values shows that the conclusion is not tautologically implied by the premise and so cannot be derived from it.

To take another example, suppose we want to show that $\sim P$ cannot be derived from the premises $[P \supset (Q \supset R)]$, $(R \supset Q)$, $(P \supset \sim Q)$. We assign truth values to P, Q, and R in such a way as to make these premises all true and the conclusion false. To make the conclusion, $\sim P$, false, we must make P true. Then, to make the third premise true, we must make Q false. But then, Q being false, in order to make the second premise true, we must make R false. This assignment of truth values (P true, Q false, R false) also makes the first premise true. Thus it is an assignment that makes the premises all true, and the conclusion false, demonstrating that the conclusion cannot be derived from the premises.

II.10 EXERCISES

A. In each of the following purported PC derivations, check the lines for errors, and indicate the numbers of the lines on which errors occur. An error is said to occur in a line if *that* line cannot be introduced in accordance with the rules of inference. This means that in checking a line for errors, you only look at that line to see whether it has the necessary relation to the previous lines, and treat the previous lines as if they were all correct. If a line is incorrect, try to find a way of correcting it (perhaps by adding additional lines).

1.	(1)	1.	P	P
	(2)	2.	$(P \supset Q)$	P
	(1, 2)	3.	Q	I, 1, 2 (I9)
2.		1.	$(P \vee Q)$	P
3.	(1)	1.	$(P \vee Q)$	P
	(2)	2.	$\sim Q$	P
	(1, 2)	3.	P	I, 1, 2 (I11)
	(1)	4.	$[\sim Q \supset (P \vee Q)]$	C, 2, 1
	(1, 2)	5.	$\{P \& [\sim Q \supset (P \vee Q)]\}$	I, 3, 4 (I13)
4.	(1)	1.	$\sim(P \& \sim Q)$	P
	(1)	2.	$(P \supset Q)$	R, 1 (E5)
	(2)	3.	$(Q \supset R)$	P
	(1, 2)	4.	$(P \supset R)$	I, 2, 3 (I12)

5.

(1)	1.	$[(R \vee S) \equiv \sim T]$	P
(1)	2.	$\{[(R \vee S) \& \sim T] \vee [\sim(R \vee S) \& \sim\sim T]\}$	R, 1 (E8)
(1)	3.	$\{[\sim T \& (R \vee S)] \vee [\sim\sim T \& \sim(R \vee S)]\}$	R, 2 (E11)
(1)	4.	$[\sim T \equiv (R \vee S)]$	R, 3 (E8)

6.

(1)	1.	$[R \equiv \sim(S \vee \sim T)]$	P
(2)	2.	$\sim R$	P
(1)	3.	$\{[R \supset \sim(S \vee \sim T)] \& [\sim(S \vee \sim T) \supset R]\}$	R, 1 (E7)
(1)	4.	$[R \supset \sim(S \vee \sim T)]$	I, 3 (I1)
(1, 2)	5.	$\sim\sim(S \vee \sim T)$	I, 2, 4 (I10)
(1, 2)	6.	$(S \vee \sim T)$	R, 5 (E1)
(7)	7.	T	P
(1, 2)	8.	$(\sim T \vee S)$	R, 6 (E12)
(1, 2, 7)	9.	S	I, 7, 8 (I11)
(1, 7)	10.	$(\sim R \supset S)$	C, 2, 9
(11)	11.	$(R \equiv S)$	P
(11)	12.	$[(R \supset S) \& (S \supset R)]$	R, 11 (E7)
(11)	13.	$(S \supset R)$	I, 12 (I2)
(1, 2, 7, 11)	14.	R	I, 9, 13 (I9)
(1, 7, 11)	15.	$(R \& \sim R)$	I, 2, 14 (I13)
(7, 11)	16.	$\{[R \equiv \sim(S \vee \sim T)] \supset (R \& \sim R)\}$	C, 1, 15
(17)	17.	R	P
	18.	$(R \supset R)$	C, 17, 17
	19.	$\sim(R \& \sim R)$	R, 18 (E5)
(7, 11)	20.	$\sim[R \equiv \sim(S \vee \sim T)]$	I, 16, 19 (I10)
(11)	21.	$\{T \supset \sim[R \equiv \sim(S \vee \sim T)]\}$	C, 7, 20
(11)	22.	$(S \supset \{T \supset \sim[R \equiv \sim(S \vee \sim T)]\})$	C, 9, 21
(11)	23.	$\{(S \& T) \supset \sim[R \equiv \sim(S \vee \sim T)]\}$	R, 22 (E15)
(11)	24.	$\{[R \equiv \sim(S \vee \sim T)] \supset \sim(S \& T)\}$	R, 23 (E16)

B. For each of the following, give a PC derivation of the formula below the line from the formulas above the line. Keep in mind the strategies for constructing PC derivations.

A. Conditionals

1. $[P \supset (Q \supset R)]$

   $[(P \supset Q) \supset (P \supset R)]$

2. $[P \supset (\sim Q \supset \sim P)]$

   $\sim(P \supset Q)$

   $\sim P$

B. Conjunctions

1. $$\begin{array}{l} (P \supset Q) \\ [P \supset (Q \supset P)] \\ \hline [(P \supset Q) \,\&\, (P \supset P)] \end{array}$$

2. $$\begin{array}{l} \{P \supset [(Q \supset R) \,\&\, (R \supset \sim S)]\} \\ (\sim P \supset P) \\ \hline (S \supset \sim R) \end{array}$$

C. Disjunctions

1. $$\begin{array}{l} P \\ \hline [(P \,\&\, Q) \lor (P \,\&\, \sim Q)] \end{array}$$

2. $$\begin{array}{l} (\sim P \supset R) \\ [\sim Q \supset (R \supset S)] \\ \hline [(P \lor Q) \lor (R \,\&\, S)] \end{array}$$

3. $$\begin{array}{l} [(R \supset S) \lor (R \,\&\, \sim T)] \\ [R \,\&\, (T \equiv \sim S)] \\ \hline \sim T \end{array}$$

D. Negations

1. $$\begin{array}{l} \sim (R \supset S) \\ [R \supset (S \lor \sim T)] \\ \hline \sim T \end{array}$$

2. $$\begin{array}{l} \sim [R \lor (\sim S \supset \sim T)] \\ \sim [(R \supset S) \supset \sim (\sim R \supset \sim Q)] \\ \hline \sim Q \end{array}$$

3. $$\begin{array}{l} [P \supset (Q \supset R)] \\ (P \supset Q) \\ (R \supset \sim Q) \\ \hline \sim P \end{array}$$

E. Biconditionals

1. $$\begin{array}{l} (P \equiv \sim Q) \\ \hline \sim (P \equiv Q) \end{array}$$

2. $$\begin{array}{l} (P \equiv Q) \\ (\sim P \equiv R) \\ \hline (Q \equiv \sim R) \end{array}$$

F. General

1. $$\begin{array}{l} (P \supset Q) \\ (\sim R \supset \sim Q) \\ \hline (P \supset R) \end{array}$$

2. $$\begin{array}{l} (P \supset Q) \\ (\sim P \supset R) \\ \hline (\sim R \supset Q) \end{array}$$

3. $$\begin{array}{l} [P \supset (Q \supset R)] \\ Q \\ \hline (P \supset R) \end{array}$$

4. $$\begin{array}{l} [P \supset (Q \supset R)] \\ \sim R \\ \hline (P \supset \sim Q) \end{array}$$

5. $$\begin{array}{l} (S \supset \sim S) \\ (R \lor S) \\ (P \supset \sim R) \\ \hline [(P \supset S) \,\&\, \sim S] \end{array}$$

6. $$\begin{array}{l} [P \supset \sim (S \,\&\, \sim T)] \\ (\sim P \supset \sim S) \\ \sim T \\ \hline \sim S \end{array}$$

7. $[(P \vee Q) \vee R]$
$[P \supset (S \equiv {\sim}T)]$
$[Q \supset {\sim}(S \vee {\sim}T)]$
$[T \supset {\sim}(R \,\&\, S)]$
$[{\sim}T \supset (R \,\&\, S)]$
———————
$(S \equiv {\sim}T)$

8. ${\sim}\{P \supset [{\sim}Q \,\&\, {\sim}(R \vee {\sim}S)]\}$
———————
$\{[P \,\&\, (R \vee {\sim}S)] \vee (P \,\&\, Q)\}$

9. $[P \supset (Q \,\&\, {\sim}Q)]$
———————
$(P \supset R)$

C. Show that the following formulas are tautologies by giving PC derivations of them; that is, derivations the last lines of which have no premise numbers.

1. $(P \vee {\sim}P)$
2. ${\sim}(P \,\&\, {\sim}P)$
3. $[(P \,\&\, Q) \supset ({\sim}Q \supset R)]$
4. $[(P \equiv Q) \supset ({\sim}P \equiv {\sim}Q)]$
5. $\{(P \supset Q) \supset [(Q \supset R) \supset (P \supset R)]\}$
6. $[P \supset ({\sim}P \supset Q)]$
7. $(P \supset P)$
8. $[P \supset (Q \supset P)]$
9. $\{[(P \,\&\, Q) \supset R] \equiv [(P \,\&\, {\sim}R) \supset {\sim}Q]\}$
10. $\{P \equiv [{\sim}P \supset (Q \,\&\, {\sim}Q)]\}$
11. $[(P \,\&\, {\sim}P) \supset Q]$

D. For each of the following, if the formula below the line is tautologically implied by the formulas above the line, construct a PC derivation to show that it is. If the formula below the line is not tautologically implied by the formulas above the line, show that it is not by a suitable assignment of truth values to the atomic parts of the formulas.

1. $[P \supset ({\sim}Q \supset R)]$
${\sim}R$
———————
$(P \supset Q)$

2. $[P \supset ({\sim}Q \supset R)]$
Q
———————
${\sim}(P \supset R)$

3. $[P \equiv (R \vee {\sim}P)]$
———————
R

4. $[P \equiv (R \vee {\sim}P)]$
———————
${\sim}R$

E. Symbolize each of the following arguments and construct a PC derivation of the conclusion from the premises:

1. If John comes to the party, Joe will not come. Joe will not come unless Mary comes, but Mary will only come if John comes. Therefore, Joe will not come to the party.

2. If the bank rejects his loan application, Mr. Horner will either have to get the money somewhere else or sell his business. If the bank will not lend him the money, he will not be able to get it anywhere else either. But if he sells his business, his only alternative will be to join the Peace Corps. Thus if the bank rejects his loan application, Mr. Horner will join the Peace Corps.
3. If the Russians beat the Americans to the moon, they will score a great propaganda victory. If the Russians announce that they are cutting their budget for space research, Congress will cut the American budget for space research. If Congress cuts the American budget for space research, but the Russians do not cut theirs, then the Russians will beat the Americans to the moon. The Russians are very devious, and will announce that they have cut their budget for space research without actually doing so. Therefore, the Russians will score a great propaganda victory.
4. If the class does well, and the professor knows it, he will not give them a test. But the professor will not know that the class does well unless he gives them a test. Therefore, if the class does well the professor will not know it.
5. If de Gaulle remains President of France, England will never get into the Common Market. If England does not get into the Common Market, she will have to trade heavily with the United States to avoid economic disaster. If England trades heavily with the United States and does not get into the Common Market, many large corporations will move out of England. If many large corporations move out of England, she will not avoid economic disaster. If England suffers an economic disaster, she will be taken over by the Communists. Consequently, if de Gaulle remains President of France, England will be taken over by the Communists.

F*. Show that we could, if we wished, adopt the following strengthened rule of tautological replacement, Rule R*, in place of Rule R without enlarging the set of theorems that can be proven or the set of formulas derivable from other formulas:

If P and Q are any formulas of the propositional calculus, and either (a) P and Q are tautologically equivalent by one of the equivalences E1–E21, or (b) we have previously constructed a derivation of either $(P \equiv Q)$ or $(Q \equiv P)$ (thus showing that P and Q are tautologically equivalent); then if any formula containing P appears on a line of a PC derivation, and a second formula can be obtained from this first formula by replacing one occurrence of P by Q, then we can write the second formula on any later line of the PC derivation, taking for premise numbers all of the premise numbers of the first formula.

G*. Show that we could eliminate all implications but I1, I2, and I3, and all equivalences but E1, E2, E4, E7, and E17, from our lists of implications and equivalences without diminishing the set of theorems that can be proven or the set of formulas derivable from other formulas. In order to prove this, establish: (1) that the result proven in question F concerning Rule R* still holds with this smaller list of implications and equivalences; and (2) that all of the implications and equivalences dropped from the lists can be established using the remaining three implications and equivalences.

II.11 THE SOUNDNESS OF PC-DERIVATIONS

In Section II.10 we assumed without proof that all theorems of the propositional calculus were tautologies, and that whenever there is a derivation of a formula from a set of formulas, the formula is tautologically implied by the set of formulas. This must now be proven. Let us say that a PC derivation is *sound* if each line is tautologically implied by its premises if it has any, and is a tautology if it does not have any. The objective of this section is to show that all PC derivations are sound.

It will be shown that if we are given any sound PC derivation, and then use one of our four rules of inference to add an additional line to it, the resulting PC derivation will still be sound. Thus the use of our rules of inference can never make a previously sound derivation unsound. This in turn implies that all PC derivations are sound, for the following reasons: The first line of a PC derivation must be a premise, because none of the other rules is applicable until we have at least one earlier line with which to work. The way we have set things up, a premise is its own only premise (that is, its own line number is its only premise number), and so a premise is tautologically implied by its premises. Thus the PC derivation consisting of just the first line of a longer PC derivation will be sound. Then, given that the use of any of our four rules to add a new line to a sound derivation yields a new derivation that is also sound, the addition of the second line of a PC derivation must also yield a sound derivation; then the addition of the third line must yield another sound derivation because the first two lines constituted a sound derivation, and so on. In other words, in a PC derivation, the first line constitutes a sound derivation, and the addition of each successive line yields a sound derivation because the previous lines constituted a sound derivation. Thus any derivation that can be constructed by successive applications of our rules of inference must be sound. Therefore, *all PC derivations are sound*; that is, every line of a PC derivation is tautologically implied by its premises if it has any, and is a tautology if it has none.[7]

Now we must go back and show for each of our rules that, if we are given any sound PC derivation, and then use that rule to add an additional line to it, the resulting PC derivation will still be sound. Let us consider our rules one at a time:

Premise introduction. We already noted that a premise is tautologically implied by its premise (which is itself), and thus if we use Rule P to add a premise to a derivation that is already sound, the new derivation will still be sound.

Rule I. We want to show that if we are given a sound PC derivation and then use Rule I to add a new line, which is implied by a set of formulas appearing earlier in the derivation, then this new line is implied by its premises if it has any or is a tautology if it has none. So suppose we have a set of formulas, all of which are tautologically implied by a certain set of premises, and suppose that this set of formulas in turn tautologically implies a further formula. Then by the strong transitivity of tautological implication, the latter formula is also tautologically implied by the set of premises. If Rule I is used to introduce this latter formula on the next line of the derivation, that line will also be implied by its premises, and so the resulting derivation will still be sound. Therefore, the use of Rule I to add a line having premises preserves the soundness of a PC derivation.

Next we must show that if we use Rule I to obtain a formula from a set of formulas, none of which has any premise numbers (hence all are tautologies), the further formula that we obtain will also be a tautology. This is because, as the further formula is tautologically implied by the set of formulas, any assignment of truth values making them all true will make the further formulas true too. But since the formulas in the set are tautologies, every assignment of truth values will make them all true. Hence every assignment of truth values will make the further formula true too, and so it is a tautology. Therefore, we can say in general that if we start with a sound PC derivation and add a new line using Rule I, the new derivation will still be sound.

Tautological Replacement. If one line is obtained from a previous line by Rule R, then the two lines are tautologically equivalent (recall the

[7] The student with a background in mathematics may note that this is a proof by mathematical induction on the number of lines in the derivation.

discussion of tautological replacement in Section II.9), and so if the first line is implied by its premises, the new line is also implied by those same premises; and if the first line is a tautology the new line is also a tautology.

Conditionalization. This rule results from the fact, which we have already seen, that if we have a set of formulas, say $R_1, \ldots, R_n, P$, which tautologically implies some further formula Q, then $R_1, \ldots, R_n \models (P \supset Q)$. Applying this now to Rule C, suppose some formula Q appears on a line of a PC derivation that is sound (and so Q is implied by its premises), and suppose that P is among the premises of Q but is not the only premise of Q. Then the set of all of those premises except P will imply $(P \supset Q)$, and so $(P \supset Q)$ will be implied by the premises that we give it using Rule C. Thus this use of Rule C will transform one sound derivation into another.

Furthermore, if in applying conditionalization we end with a line having no premise numbers, then the formula on that line will be a tautology. This is because, given that the original derivation was sound, the antecedent of the conditional we obtain must have been the single premise of the line on which the consequent appeared, and so must have tautologically implied the consequent. But then the conditional itself must be a tautology, by the definition of tautological implication.

Therefore, all of the rules of inference of the propositional calculus have the characteristic that if they are used to add a line to a PC derivation that is sound, the resulting PC derivation will still be sound. And, as we have seen, this entails that all PC derivations are sound; that is, that *every line of a PC derivation is tautologically implied by its premises if it has any, and is a tautology if it does not have any*. In particular, *every theorem of the propositional calculus is a tautology*. Thus PC derivations give us another way to establish that formulas are tautologies or tautologically imply one another.

We have seen that all theorems of the propositional calculus are tautologies, and thus that PC derivations give us a way other than truth tables to establish that a formula is a tautology. But this is not to say that *all* tautologies are theorems of the propositional calculus. It might be the case that using PC derivations we can only derive some restricted class of all the formulas that are tautologies. If this were true, then we would still need truth tables to show that the rest of the tautologies were tautologies. Fortunately, this is not the case. In Section II.12 it will be shown that all tautologies are theorems of the propositional calculus, and that PC derivations constitute a genuine alternative to truth tables. But there is still a big difference between PC derivations and truth tables. The construction of truth tables is purely mechanical and requires no ingenuity. You just insert the formulas and grind out T's and F's according to the rules. On the other hand, the construction of a PC derivation of a formula may require considerable ingenuity. PC derivations have two advantages over truth tables. The first is that when you can see how to construct a derivation of a given formula it will often be much easier than writing out a truth table. But if ingenuity fails you, then truth tables will be easier. Thus PC derivations provide a shorter way of establishing that formulas are tautologies, but only at the expense of being nonmechanical and requiring ingenuity for their construction.[8] The second, and perhaps more important advantage, is that PC derivations do more than just provide a way of establishing that a formula is a tautology. PC derivations also reflect our natural nonformal reasoning; thus familiarity with PC derivations will givc insight into the natural patterns of inference that we normally use, and make it much easier to tell at a glance that complicated arguments are sound.

II.12 BOOLEAN NORMAL FORM AND COMPLETENESS

The principal aim of this section is to show that all tautologies are theorems of the propositional calculus. We will express this by saying that PC derivations are *complete*. We have already seen that the converse is true, namely, that all theorems of the propositional calculus are tautologies. In order to prove completeness, our strategy will be as follows. First we will introduce the concept of a formula being in *Boolean normal form*. A formula in Boolean normal form will be any formula having a certain very simple form that will be described below. Then we will show the following:

1. Every formula is tautologically equivalent to a formula in Boolean normal form. The latter formula is said to be obtained by "reducing" the former to Boolean normal form.

[8] Actually, mechanical rules can be given for constructing PC derivations so that this requires no more ingenuity than the construction of truth tables. But the construction of PC derivations using those mechanical rules will generally be no shorter, and frequently a good deal longer, than the construction of a truth table. So the mechanical rules are not really of much use. It is only through ingenuity that we can shorten the process of establishing that formulas are tautologies.

2. Every formula can be derived from the formula we get by reducing it to Boolean normal form.
3. Every formula in Boolean normal form that is a tautology is a theorem of the propositional calculus.

Having established these three conclusions, we can then reason as follows. If a formula is a tautology, then the formula we obtain by reducing it to Boolean normal form is also a tautology, because they are tautologically equivalent (conclusion 1). The latter formula is a theorem of the propositional calculus (conclusion 3). So let us take the derivation of the formula in Boolean normal form, and then add to it the derivation of the original tautology from the formula in Boolean normal form. The latter derivation must exist by conclusion 2. This gives us a derivation of the original tautology, thus showing that it is a theorem of the propositional calculus. Therefore, every tautology is a theorem of the propositional calculus.

Now what remains is to establish conclusions 1, 2, and 3. First we need the concept of a *Boolean conjunction* of a finite set of atomic formulas. This is most easily explained with an example. Given the three atomic formulas, P, Q, and R, the Boolean conjunctions of these formulas are the following: $(P \& Q \& R)$, $(P \& Q \& \sim R)$, $(P \& \sim Q \& R)$, $(P \& \sim Q \& \sim R)$, $(\sim P \& Q \& R)$, $(\sim P \& Q \& \sim R)$, $(\sim P \& \sim Q \& R)$, $(\sim P \& \sim Q \& \sim R)$. In other words, the Boolean conjunctions of P, Q, and R are the conjunctions of three formulas such that, for each P, Q, and R, either it or its negation (but not both) is a conjunct, and the conjuncts occur in alphabetical order. Analogously, we define the general concept of a Boolean conjunction as follows: Given a finite list of atomic formulas, a Boolean conjunction of those formulas is any conjunction such that (1) each of its conjuncts is either one of the atomic formulas in the list or the negation of one of them, (2) for each atomic formula in the list, either it or its negation (but not both) is a conjunct of the conjunction, and (3) all of the conjuncts occur in alphabetical order.[9]

In the degenerate case where we have a list of just one atomic formula, for example, P, we say that there are two Boolean conjunctions—the atomic formula itself, and its negation (such as, P and

[9] As some of the atomic formulas may have subscripts, this complicates what we mean by alphabetical order. By "alphabetical order" we mean the ordering starting from A and going through Z, then A_1 through Z_1, then A_2 through Z_2, and so on.

$\sim P$). These are not conjunctions, but we call them Boolean conjunctions anyway.

The Boolean conjunctions of a set of atomic formulas correspond to lines of a truth table of a formula having those atomic formulas as its atomic parts. For example, we get the following correspondence between the Boolean conjuncts of P, Q, and R, and the left side of a truth table of a formula whose atomic parts are P, Q, and R:

	P	Q	R	. . .
$(P \,\&\, Q \,\&\, R)$	T	T	T	.
$(P \,\&\, Q \,\&\, \sim R)$	T	T	F	.
$(P \,\&\, \sim Q \,\&\, R)$	T	F	T	.
$(P \,\&\, \sim Q \,\&\, \sim R)$	T	F	F	.
$(\sim P \,\&\, Q \,\&\, R)$	F	T	T	.
$(\sim P \,\&\, Q \,\&\, \sim R)$	F	T	F	.
$(\sim P \,\&\, \sim Q \,\&\, R)$	F	F	T	.
$(\sim P \,\&\, \sim Q \,\&\, \sim R)$	F	F	F	.

For each Boolean conjunction, there is one and only one line of the truth table on which it is true, so we can correlate it with that line. We can think of a Boolean conjunction as listing the truth values its atomic parts must have in order for it to be true; for example, $(\sim P \,\&\, Q \,\&\, \sim R)$ says that P is false, Q is true, and R is false.

Next we want the concept of a *formula in Boolean normal form.* We say that a formula of the propositional calculus is in Boolean normal form if, and only if, it is either a Boolean conjunction or it is a disjunction of some of the Boolean conjunctions of its atomic parts. So, for example, the following are all formulas in Boolean normal form:

$[(P \,\&\, Q) \vee (\sim P \,\&\, \sim Q)]$
$[(P \,\&\, Q) \vee (P \,\&\, \sim Q) \vee (\sim P \,\&\, Q) \vee (\sim P \,\&\, \sim Q)]$
$(P \,\&\, Q)$

On the other hand, the following formulas are not in Boolean normal form:

$[(P \,\&\, Q) \vee (\sim P \,\&\, R)]$
$[P \vee (\sim P \,\&\, Q)]$

The disjuncts of these latter formulas are not Boolean conjunctions of the atomic parts of these formulas, because at least one of the disjuncts in each fails to contain an occurrence of all of the atomic parts of the formula.

What makes formulas in Boolean normal form interesting is that we can tell just by looking at them whether they are tautologies. In general, a formula in Boolean normal form is a tautology if, and only if, it is the disjunction of *all* of the Boolean conjunctions of its atomic parts. Thus, the second example above of a formula in Boolean normal form is a tautology, but the others are not. Why is this so? Recall first that a Boolean conjunction is true on one and only one line of its truth table, and in effect is just a list of the truth values the atomic formulas have on that line. A formula in Boolean normal form is a disjunction of Boolean conjunctions. A disjunction is true if, and only if, at least one of its disjuncts is true, so a formula in Boolean normal form is true if, and only if, at least one of the Boolean conjunctions in it is true. Thus the formula in Boolean normal form is true on just those lines that correspond to the Boolean conjunctions that make it up. We can think of a formula in Boolean normal form as simply listing those lines of the truth table on which it is true. For example, the formula $[(P \,\&\, Q) \vee (\sim P \,\&\, \sim Q)]$ is true on just those lines on which either $(P \,\&\, Q)$ is true or $(\sim P \,\&\, \sim Q)$ is true; that is, on the line on which P is T and Q is T, and the line on which P is F and Q is F.

If a formula in Boolean normal form is the disjunction of *all* of the Boolean conjunctions of its atomic parts, then it is true on every line of the truth table (because each line will make one of the Boolean conjunctions true); therefore it is a tautology. On the other hand, if a formula in Boolean normal form is not the disjunction of all of the Boolean conjunctions of its atomic parts, then all of its disjuncts will be false on those lines of the truth table that correspond to the omitted Boolean conjunctions (because each Boolean conjunction is true only on the line corresponding to it); thus the formula itself will be false on those lines. Therefore a formula in Boolean normal form that is not the disjunction of all of the Boolean conjunctions of its atomic parts is not a tautology. *A formula in Boolean normal form is a tautology if, and only if, it is the disjunction of all of the Boolean conjunctions of its atomic parts.*

The next thing to be shown is that if a formula in Boolean normal form is a tautology, then it is also a theorem of the propositional calculus. To prove this we first note that $(P \vee \sim P)$ is a theorem of the propositional calculus. We can give the following derivation of $(P \vee \sim P)$:

(1)	1.	$\sim P$	P
	2.	$(\sim P \supset \sim P)$	C, 1
	3.	$(\sim\sim P \vee \sim P)$	R, 2 (E4)
	4.	$(P \vee \sim P)$	R, 3 (E1)

This use of Rule C made on line 2, where we take both the antecedent and consequent of the conditional from the same line, is rather unusual, but it is in accordance with the rule. Notice also that any substitution instance of $(P \lor \sim P)$ that results from substituting some other atomic formula for P is also a theorem of the propositional calculus. We can, for example, construct a derivation of $(Q \lor \sim Q)$ just like the derivation of $(P \lor \sim P)$ except that we replace P by Q throughout.

The next thing to observe is that any conjunction of substitution instances of $(P \lor \sim P)$, which result from replacing P by other atomic formulas, is also a theorem of the propositional calculus; for example, $[(P \lor \sim P) \& (Q \lor \sim Q)]$ is a theorem of the propositional calculus. To construct a derivation of this we simply let the first four lines of the derivation be the derivation of $(P \lor \sim P)$, the next four lines be the derivation of $(Q \lor \sim Q)$, and then on the next line we use Rule I and I15 to get $[(P \lor \sim P) \& (Q \lor \sim Q)]$. Similarly we can get a derivation of $[(P \lor \sim P) \& (Q \lor \sim Q) \& (R \lor \sim R)]$, and so forth. All of these formulas are theorems of the propositional calculus.

Now we must establish a general result about the way conjunctions of disjunctions combinc. Suppose we have n different formulas, $P_1, \ldots, P_n$, and two other formulas, R and S. What we want to show is that from the formula $[(R \lor S) \& (P_1 \lor P_2 \lor \ldots \lor P_n)]$ we can derive the consequence

$$[(P_1 \& R) \lor (P_1 \& S) \lor (P_2 \& R) \lor (P_2 \& S) \lor \ldots \lor (P_n \& R) \lor (P_n \& S)]$$

The derivation will go as follows:

(1)	1.	$[(R \lor S) \& (P_1 \lor \ldots \lor P_n)]$	P
(1)	2.	$[((R \lor S) \& P_1) \lor ((R \lor S) \& (P_2 \lor \ldots \lor P_n))]$	R, 1 (E13)
(1)	3.	$[((R \lor S) \& P_1) \lor ((R \lor S) \& P_2) \lor ((R \lor S) \& (P_3 \lor \ldots \lor P_n))]$	R, 2 (E13)
	⋮	(repeated application of R with E13)	
(1)	n	$[((R \lor S) \& P_1) \lor \ldots \lor ((R \lor S) \& P_n)]$	R, n (E13)
(1)	$n+1$	$[(P_1 \& (R \lor S)) \lor ((R \lor S) \& P_2) \lor \ldots \lor ((R \lor S) \& P_n)]$	R, n (E11)
(1)	$n+2$	$[(P_1 \& (R \lor S)) \lor (P_2 \& (R \lor S)) \lor ((R \lor S) \& P_3) \lor \ldots \lor ((R \lor S) \& P_n)]$	R, $n+1$ (E11)
	⋮	(repeated application of R with E11)	

(1) $2n$ $[(P_1 \& (R \vee S))$
$\vee \ldots \vee (P_n \& (R \vee S))]$ R, $2n - 1$ (E11)

(1) $2n + 1$ $[(P_1 \& R) \vee (P_1 \& S)$
$\vee (P_2 \& (R \vee S))$
$\vee \ldots \vee (P_n \& (R \vee S))]$ R, $2n$ (E13)

⋮ (repeated application of R with E13)

(1) $3n$ $[(P_1 \& R) \vee (P_1 \& S)$
$\vee \ldots \vee (P_n \& R) \vee (P_n \& S)]$ R, $3n - 1$ (E13)

In this derivation, we begin by taking $[(R \vee S) \& (P_1 \vee \ldots \vee P_n)]$ as a premise. Then we use Rule R with E13 to bring $(R \vee S)$ inside the disjunction $(P_1 \vee \ldots \vee P_n)$. After repeating this process n-1 times, we have derived $[((R \vee S) \& P_1) \vee \ldots \vee ((R \vee S) \& P_n)]$. Then we use Rule R with E11 n times to turn each of these disjuncts around, giving us the disjunction $[(P_1 \& (R \vee S)) \vee \ldots \vee (P_n \& (R \vee S))]$. Then we use Rule R and E13 again to expand each disjunct of the form $(P_i \& (R \vee S))$ to $[(P_i \& R) \vee (P_i \& S)]$. This gives us the consequence: $[(P_1 \& R) \vee (P_1 \& S) \vee \ldots \vee (P_n \& R) \vee (P_n \& S)]$, which can be derived from the premise.

In order to make the above general argument a bit clearer, let us illustrate it with a particular case. Letting $n = 3$, suppose we want to derive

$$[(P_1 \& R) \vee (P_1 \& S) \vee (P_2 \& R) \vee (P_2 \& S) \vee (P_3 \& R) \vee (P_3 \& S)]$$

from $[(R \vee S) \& (P_1 \vee P_2 \vee P_3)]$. Then the above derivation would go as follows:

(1) 1. $[(R \vee S) \& (P_1 \vee P_2 \vee P_3)]$ P

Steps (2)–(n)

(1) 2. $[((R \vee S) \& P_1) \vee ((R \vee S) \& (P_2 \vee P_3))]$ R, 1 (E13)

(1) 3. $[((R \vee S) \& P_1) \vee ((R \vee S) \& P_2)$
$\vee ((R \vee S) \& P_3)]$ R, 2 (E13)

Steps ($n + 1$)–($2n$)

(1) 4. $[(P_1 \& (R \vee S)) \vee ((R \vee S) \& P_2)$
$\vee ((R \vee S) \& P_3)]$ R, 3 (E11)

(1) 5. $[(P_1 \& (R \vee S)) \vee (P_2 \& (R \vee S))$
$\vee ((R \vee S) \& P_3)]$ R, 4 (E11)

(1) 6. $[(P_1 \& (R \vee S)) \vee (P_2 \& (R \vee S))$
$\vee (P_3 \& (R \vee S))]$ R, 5 (E11)

Steps $(2n + 1)$–$(3n)$

(1) 7. $[(P_1 \& R) \vee (P_1 \& S) \vee (P_2 \& (R \vee S))$
$\vee (P_3 \& (R \vee S))]$ R, 6 (R13)

(1) 8. $[(P_1 \& R) \vee (P_1 \& S) \vee (P_2 \& R)$
$\vee (P_2 \& S) \vee (P_3 \& R) \vee (P_3 \& S)]$ R, 8 (E13)

(1) 9. $[(P_1 \& R) \vee (P_1 \& S) \vee (P_2 \& R)$
$\vee (P_2 \& S) \vee (P_3 \& R) \vee (P_3 \& S)]$ R, 8 (E13)

This result will now be used to show that if a formula in Boolean normal form is a tautology, then it is a theorem of the propositional calculus. We have already seen that any conjunction of formulas of the form $(P \vee \sim P)$ is a theorem of the propositional calculus. Now we must show that, if we are given a list of atomic formulas, then from the conjunction of all formulas of the form $(P \vee \sim P)$, which result from substituting those atomic formulas for P, we can derive the disjunction of all the Boolean conjunctions of these atomic formulas. We begin by supposing that we have a list of two atomic formulas, say P and Q, and show that the disjunction of all the Boolean conjunctions of P and Q is derivable from $[(Q \vee \sim Q) \& (P \vee \sim P)]$. The disjunction of all Boolean conjunctions of P and Q is

$$[(P \& Q) \vee (P \& \sim Q) \vee (\sim P \& Q) \vee (\sim P \& \sim Q)]$$

By the above general result, letting $n = 2$, and letting Q be R, $\sim Q$ be S, P be P_1, and $\sim P$ be P_2, we see that

$$[(P \& Q) \vee (P \& \sim Q) \vee (\sim P \& Q) \vee (\sim P \& \sim Q)]$$

is derivable from $[(Q \vee \sim Q) \& (P \vee \sim P)]$. Therefore, the disjunction of all Boolean conjunctions of two atomic formulas can be derived from the conjunction of formulas of the form $(P \vee \sim P)$ which results from substituting each of those two atomic formulas for P. Therefore, all formulas having only two atomic parts that are in Boolean normal form and are tautologies are theorems of the propositional calculus.

Next let us suppose that, for some arbitrary number k, if we are given a list of k atomic formulas, then from the conjunction of all formulas of the form $(P \vee \sim P)$, which result from substituting those atomic formulas for P, we can derive the disjunction of all the Boolean conjunctions of those k atomic formulas, and then show that it follows from this that the same thing is true if we have a list of $k + 1$ atomic formulas. If we can show this, it will follow that for any number n, if we are given n atomic formulas, we can derive the disjunction of all the Boolean conjunctions of those atomic formulas from the conjunction of all formulas of the form $(P \vee \sim P)$ that result from substituting those atomic formulas for P. This is because as we have just seen, this is true if $n = 2$. According to the above result, which we will establish shortly, if it is true for $n = 2$, then it is true for $n = 3$. But if it is true for $n = 3$ then it is true for $n = 4$, and so on. Therefore it is true for all n.

Now let us establish that if this is true for k, it is true for $k + 1$ (regardless of what number k is). Suppose that the disjunction of all Boolean conjunctions of k atomic formulas can be derived from the conjunction of all formulas of the form $(P \vee \sim P)$ that result from substituting the k atomic formulas for P. Suppose the atomic formulas are $P_1, \ldots, P_k$. Now let us add a new atomic formula P_{k+1} to the list. Then consider the conjunction

$$[(P_{k+1} \vee \sim P_{k+1}) \;\&\; (P_k \vee \sim P_k) \;\&\; \ldots \;\&\; (P_1 \vee \sim P_1)]$$

and show that from this we can derive the disjunction of all Boolean conjunctions of the new list of $k + 1$ atomic formulas. From the above conjunction we can derive $[(P_k \vee \sim P_k) \vee \ldots \vee (P_1 \vee \sim P_1)]$ using Rule I with I2. We are supposing that from this we can derive the disjunction of all Boolean conjunctions of the original list of k atomic formulas. Let us call this disjunction $[Q_1 \vee \ldots \vee Q_{2^k}]$, where Q_i is the ith (e.g., 4th) Boolean conjunction.[10] So we can derive this disjunction from our long conjunction. By Rule I and I1 we can also derive $(P_{k+1} \vee \sim P_{k+1})$ from the conjunction. Then using Rule I and I13 we can put these together, getting $[(P_{k+1} \vee \sim P_{k+1}) \;\&\; (Q_1 \vee \ldots \vee Q_{2^k})]$. Applying the general result above, where we now let R be P_{k+1}, S be $\sim P_{k+1}$, $n = 2^k$, and the P_i be Q_i, we see that from this we can derive

$$[(Q_1 \;\&\; P_{k+1}) \vee (Q_1 \;\&\; \sim P_{k+1}) \vee \ldots \vee (Q_{2^k} \;\&\; P_{k+1}) \vee (Q_{2^k} \;\&\; \sim P_{k+1})]$$

But this formula is the disjunction of all Boolean conjunctions of the $k + 1$ atomic formulas, because each such Boolean conjunction results from appending either P_{k+1} or $\sim P_{k+1}$ to the end of one of the Boolean conjunctions of the original k atomic formulas. Therefore, if we can derive the disjunctions of all Boolean conjunctions of k atomic formulas

[10] If there are k atomic formulas, then there are 2^k Boolean conjunctions, just as there are 2^k lines in the truth table.

from the conjunction of all formulas of the form $(P \vee \sim P)$, which result from substituting the k atomic formulas for P, then this is also true for $k + 1$ atomic formulas. And, as we have seen, this means that it is true for any number of atomic formulas.

Therefore, given any list of atomic formulas, the disjunction of all Boolean conjunctions of those atomic formulas can be derived from the conjunction of all formulas of the form $(P \vee \sim P)$ that result from substituting the atomic formulas for P. But we have seen that this latter conjunction is a theorem of the propositional calculus. So we can take the derivation of this latter formula, add to the end of it the derivation of the disjunction of the Boolean conjunctions, and have a derivation of it. Therefore, the disjunction of all of the Boolean conjunctions is itself a theorem of the propositional calculus. And since these are the only formulas in Boolean normal form that are tautologies, it follows that *all formulas in disjunctive normal form that are tautologies are theorems of the propositional calculus.*

Having shown that all tautologies that are in Boolean normal form are theorems of the propositional calculus, we can go on to show that *all* tautologies are theorems of the propositional calculus. This will be done by showing that every formula of the propositional calculus can be reduced to a formula in Boolean normal form to which it is tautologically equivalent. This process of reduction will be called *reduction to Boolean normal form,* and it can be carried out in a PC derivation in such a way that it will also be true that any formula can be derived from the formula that results from reducing it to Boolean normal form.

We begin by looking at a particular formula and seeing how it can be reduced to Boolean normal form. Consider

$$\{P \supset [(P \supset Q) \,\&\, \sim(\sim Q \vee \sim P)]\}$$

The first thing we do is use Rule R and E4 and E8 to replace all conditionals and biconditionals by disjunctions. In this case there are no biconditionals. So using E4 we can first reduce the above formula to

$$\{\sim P \vee [(\sim P \vee Q) \,\&\, \sim(\sim Q \vee \sim P)]\}$$

We have now eliminated all conditionals and biconditionals, so that the only sentential connectives remaining in the formula are "&," "$\vee$," and "$\sim$."

The next step is to drive the negation signs inwards until they occur only in front of atomic formulas. To do this we use E1, E2, and E3. We begin with the outermost negation sign and work inwards. In this case, this first gives us

$$\{{\sim}P \vee [({\sim}P \vee Q) \,\&\, ({\sim}{\sim}Q \,\&\, {\sim}{\sim}P)]\}$$

and then

$$\{{\sim}P \vee [({\sim}P \vee Q) \,\&\, (Q \,\&\, P)]\}$$

We have now driven the negation signs in so that the only formulas that are negated are atomic formulas.

The next step is to expand the formula to a disjunction of conjunctions using Rule R and E13 and E11. Whenever we have a conjunction one of whose conjuncts is a disjunction, we use E11 to get it in the right order and then E13 to transform it into a disjunction of two conjunctions. First, using E11 to reverse the order of the conjuncts so that we can later use E13, we get

$$\{{\sim}P \vee [(Q \,\&\, P) \,\&\, ({\sim}P \vee Q)]\}$$

and then, using E13, we get

$$\{{\sim}P \vee [((Q \,\&\, P) \,\&\, {\sim}P) \vee ((Q \,\&\, P) \,\&\, Q)]\}$$

Now our formula is almost in disjunctive normal form. It is a long disjunction of conjunctions. The only difficulty is that the conjunctions are not Boolean conjunctions because (1) the first one $({\sim}P)$ does not contain an occurrence of Q; (2) the second and third disjuncts contain more than one occurrence each of P and of Q; and (3) P and Q do not appear in alphabetical order in the conjunctions. We will remedy these difficulties one at a time.

First, ${\sim}P$ is not a Boolean conjunction because it contains no occurrence of Q. This is remedied by using Rule R with E20. By this equivalence, our formula is equivalent to

$$\{[{\sim}P \,\&\, (Q \vee {\sim}Q)] \vee [((Q \,\&\, P) \,\&\, {\sim}P) \vee ((Q \,\&\, P) \,\&\, Q)]\}$$

Now, expanding the formula with E13 again, we get

$$\{[({\sim}P \,\&\, Q) \vee ({\sim}P \,\&\, {\sim}Q)] \\ \vee [((Q \,\&\, P) \,\&\, {\sim}P) \vee ((Q \,\&\, P) \,\&\, Q)]\}$$

By using this technique we can always introduce new atomic parts into conjunctions that lack them.

Secondly, $((Q \& P) \& \sim P)$ and $((Q \& P) \& Q)$ contain more than one occurrence of P and of Q respectively. To remedy this we first use Rule R with E9 and E11 to get the double occurrences together:

$$\{[(\sim P \& Q) \vee (\sim P \& \sim Q)] \\ \vee [((Q \& P) \& \sim P) \vee (Q \& (Q \& P))]\}$$

$$\{[(\sim P \& Q) \vee (\sim P \& \sim Q)] \\ \vee [(Q \& (P \& \sim P)) \vee ((Q \& Q) \& P)]\}$$

Now, in one of our disjuncts there is an atomic formula conjoined with itself, and in another there is an atomic formula conjoined with its negation. These are handled separately. When we have an atomic formula conjoined with itself we use E19 to eliminate one of its occurrences:

$$\{[(\sim P \& Q) \vee (\sim P \& \sim Q)] \vee [(Q \& (P \& \sim P)) \vee (Q \& P)]\}$$

Next, to get rid of an atomic formula conjoined with its negation, we first use E21 to eliminate the rest of the conjunction:

$$\{[(\sim P \& Q) \vee (\sim P \& \sim Q)] \vee [(P \& \sim P) \vee (Q \& P)]\}$$

then invert the order of the disjuncts using E11:

$$\{[(\sim P \& Q) \vee (\sim P \& \sim Q)] \vee [(Q \& P) \vee (P \& \sim P)]\}$$

and finally eliminate that disjunct altogether using E17:

$$\{[(\sim P \& Q) \vee (\sim P \& \sim Q)] \vee (Q \& P)\}$$ [11]

Now the only thing that remains is to arrange the conjuncts in alphabetical order, which is done using E11 (and E9 if necessary):

$$[(\sim P \& Q) \vee (\sim P \& \sim Q) \vee (P \& Q)]$$

This is a formula in Boolean normal form that is tautologically equivalent to our original formula.

The above process of reduction to Boolean normal form is completely general. The only time it will not work is when, upon reducing the formula to a disjunction of conjunctions of its atomic formulas and their negations, we find that every disjunct contains an atomic formula both negated and unnegated. Then we can use E21 and E17 to cross out all but one of the disjuncts; the one that remains is not a Boolean

[11] Note that this process could not be carried out if $(P \& \sim P)$ were the only disjunct; that is, the whole formula. But under any other circumstances it can be carried out.

conjunction. But notice that if a formula reduces to such a disjunction, it is false on every line of its truth table, and so its negation is a tautology. This is because none of those conjunctions can ever be true, and so no disjunct of the disjunction is every true, making the disjunction always false. However, any formula whose negation is not a tautology can be reduced to Boolean normal form by following the recipe illustrated above:

1. Replace all conditionals and biconditionals by disjunctions, using E4 and E8.
2. Drive all negations in so that the only formulas negated are atomic, using E1, E2, and E3.
3. Transform the formula into a disjunction of conjunctions of atomic formulas and their negations, using E11 and E13.
4. Expand any of the disjuncts not containing occurrences of all the atomic parts, using E20 and E13.
5. When the same atomic formula appears twice in the same disjunct, eliminate one occurrence of it by using E9, E11, and E19.
6. When the same atomic formula occurs both negated and unnegated in the same disjunct, eliminate the disjunct using E9, E11, E21, and E17.

Notice that these steps can be carried out entirely within a PC derivation by repeatedly using Rule R. Notice also that the steps are reversible so that we can also derive the original formula from the formula we get by reduction to Boolean normal form. This latter fact is what allows us to show that all tautologies are theorems of the propositional calculus.

To show that all tautologies are theorems of the propositional calculus, we reason as follows: Suppose we have a formula that is a tautology. When we reduce it to Boolean normal form, the resulting formula will also be a tautology, because it is tautologically equivalent to the original formula. As we have seen, if a formula in Boolean normal form is a tautology, then it is a theorem of the propositional calculus. So let us construct a derivation of the formula in Boolean normal form. But we have just seen that the original formula can be derived from the formula we get by reduction to Boolean normal form. Thus by adding this derivation onto the end of the derivation of the formula in Boolean normal form, we obtain a derivation of the original tautology, and so it too is a theorem of the propositional calculus. Therefore, *every tautology is a theorem of the propositional calculus*; that is, PC derivations are complete. This shows that PC derivations constitute a genuine alternative to truth tables. Any formula that can be shown to be a tautology by using truth tables can be derived in a PC derivation.

One final observation about PC derivations: because they are complete, it follows that if a formula is tautologically implied by a set of formulas, then it can be derived from them in a PC derivation. The reason for this is as follows: Suppose a formula Q is tautologically implied by the set of formulas $P_1, \ldots, P_n$. Then, by the definition of tautological implication, the conditional $[(P_1 \,\&\, \ldots \,\&\, P_n) \supset Q]$ is a tautology. Then, by the completeness of PC derivations, this conditional is a theorem of the propositional calculus. Now, we can construct a PC derivation of Q from the premises $P_1, \ldots, P_n$ as follows. First we introduce the premises $P_1, \ldots, P_n$ using Rule P. Then by Rule I, using I15, we can derive the conjunction $(P_1 \,\&\, \ldots \,\&\, P_n)$. Next we insert the derivation of $[(P_1 \,\&\, \ldots \,\&\, P_n) \supset Q]$. Then, by *modus ponens* (I9), we can derive Q. Therefore, *if a formula is tautologically implied by a set of formulas, then it can be derived from them in a PC derivation.*

II.13 EXERCISES

A. Which of the following formulas are in Boolean normal form? Of those that are, which are tautologies?

1. $(P \,\&\, Q)$
2. $[(P \,\&\, Q) \vee (\sim P \,\&\, Q) \vee (\sim P \,\&\, \sim Q) \vee (P \,\&\, \sim Q)]$
3. $[(P \,\&\, Q) \vee \sim(P \,\&\, Q)]$
4. $[(R \,\&\, \sim S) \vee (S \,\&\, \sim T)]$
5. $\{[(P \,\&\, Q) \vee (P \,\&\, \sim Q)] \supset (\sim P \,\&\, Q)\}$
6. $[(R \,\&\, \sim S) \vee S]$
7. P
8. $(P \vee \sim P)$

B. Reduce the following formulas to Boolean normal form:

1. $(P \supset Q)$
2. $[P \equiv \sim(P \,\&\, \sim Q)]$
3. $[(P \equiv Q) \equiv \sim(P \equiv \sim Q)]$
4. $(P \,\&\, Q)$
5. $[P \,\&\, (Q \vee (R \equiv \sim P))]$

C. "&," "∨," "∼," "⊃," and "≡" are called "truth-functional connectives" because the truth values of formulas constructed out of them are functions merely of the truth values of their atomic parts. We can talk about other sentential connectives too, like "neither . . . nor," "unless," "then," and so on. In general, given a connective symbolized by some symbol, say "∗," we say that the connective is *truth-functional* if, and only if, the truth value of $(P * Q)$ is a function simply of the truth values of P and Q, and so can be represented by a truth table.

1. Which of the following connectives are truth-functional? For those that are, construct truth tables.

    a. Neither *P* nor *Q*.
    b. If it were true that *P*, then it would be true that *Q*.
    c. It is false that *P*, but it is true that *Q*.
    d. *P* and then *Q*.
    e. If *P*, then necessarily *Q*.

2. Some truth functional connectives can be introduced by definition in terms of others; for example, by E4 we can define $(P \supset Q)$ as $({\sim}P \vee Q)$, and by E8 we can define $(P \equiv Q)$ as

$$[(P \supset Q) \ \& \ (Q \supset P)]$$

    a. Show that "∨," "⊃," and "≡" can all be introduced by definition using only the connectives "&" and "∼."
    b*. Using what you know about Boolean normal forms, show that every truth-functional connective "∗" that connects two formulas (like "&" and "∨") can be defined using only the connectives "&" and "∼."
    c*. Show that "∨" can be defined using only the connective "⊃."
    d*. Let $(P \mid Q)$ mean "Neither *P* nor *Q*." Show that using this single connective you can define "&" and "∼," and hence all other truth-functional connectives.

III THE PREDICATE CALCULUS

III.1 LIMITATIONS OF THE PROPOSITIONAL CALCULUS

We saw in the last chapter that we could deal with many statement forms and argument forms within the propositional calculus. But, unfortunately, there are also a number of important statement forms and argument forms with which we cannot deal in the propositional calculus. For instance, we have no way of expressing the form of statements like "All men are mortal" in the propositional calculus. The form of this statement is "All *A* are *B*," but it cannot be symbolized as a formula of the propositional calculus. And similarly, the argument form

1. All *A* are *B*.
2. All *B* are *C*.

---

3. Therefore, all *A* are *C*.

is sound, but cannot be expressed in the propositional calculus. In general, those statement forms and argument forms that involve such words as "all," "each," "every," and "some" or "there is," cannot be symbolized in the propositional calculus. It is advisable to extend our logical apparatus so that we can deal with these statement and argument forms. The logical theory that we get when we do this is called *the predicate calculus*.

III.2 PREDICATES, RELATIONS, AND INDIVIDUAL CONSTANTS

In order to talk about the forms of statements like "All men are mortal" or "All apples are red," we must introduce the concept of a *predicate*. When we say something about some individual (a person, object, number, and so on), we ascribe a predicate to that individual. If we say "John is a bachelor," we ascribe the predicate "is a bachelor" to John. In order to symbolize the form of a statement like "John is a bachelor," we use capital letters with superscript "1," and with or without numerical subscripts, to stand for predicates. We might use B^1 to stand for "is a bachelor." Such letters are called *predicate letters*.[1] We must also have some way of symbolizing those expressions, like "John," or "the girl with the red hair," that denote individuals. For this purpose we use the lower case letters from a through s, with or without numerical subscripts; they are called *individual constants*. We might let j stand for "John." Using predicate letters and individual constants we can symbolize the form of the statement "John is a bachelor" as B^1j. In other words, we write the predicate letter followed immediately by the individual constant. Whenever we have a statement of the form "b is A" (for example, "This apple is red," or "James is angry"), we symbolize it in the predicate calculus as A^1b.

We may combine predicate letters and individual constants with the sentential connectives to symbolize the forms of more complicated statements. If we want to symbolize the form of "If John is a bachelor then he is unmarried," we can let U^1 be "is unmarried," B^1 be "is a bachelor," and j be "John," and then symbolize the complex statement $(B^1j \supset U^1j)$.

In some statements we encounter *relations*; for example, the statement "John is taller than Bill" employs the relation "is taller than." In analogy to our treatment of predicates, we might symbolize this as T^2jb, letting j stand for "John," b for "Bill," and T^2 for the relation "is taller than"; T^2 here is called a *relation symbol*. Relation symbols are just capital letters with numerical superscripts, with or without numerical subscripts. They differ from predicate letters and sentential letters only in their superscripts.

[1] Note that a predicate letter is distinguished from a sentential letter only by its superscript.

Notice that it makes a difference in which order we write the individual constants after the relation symbol. If T^2jb means "John is taller than Bob," then T^2bj means "Bob is taller than John," which is something quite different. In specifying the meaning of a relation symbol, we must say in what order the individual constants are to be written. We can do this with the following device. Rather than just saying that T^2 stands for the relation "is taller than," we can say that it stands for the relation "(1) is taller than (2)." Here the numbers in parentheses tell us in what order the individual constants are to be written after T^2. If we have the statement "John is taller than Bob," we compare it with the relation "(1) is taller than (2)" and see what is in the place of the (1) and what is in the place of the (2). In this case "John" is in the place of the (1) and "Bob" is in the place of the (2); so this means that "John" comes first and "Bob" comes second. Thus we symbolize our statement as T^2jb. If instead we had taken T^2 to stand for the relation "(2) is taller than (1)," then in the statement "John is taller than Bob," "John" would occupy the place of the (2), indicating that it comes second, and "Bob" would occupy the place of the (1), indicating that it comes first. Then the statement would be symbolized as T^2bj. We might diagram this as follows:

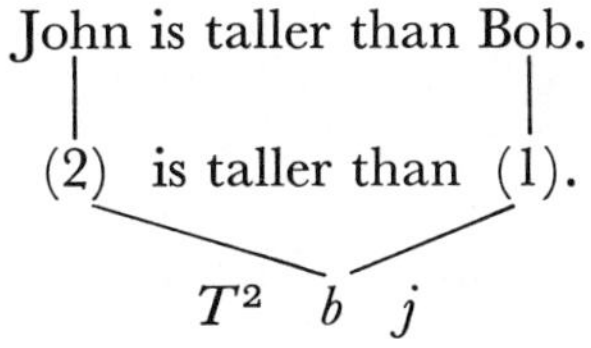

In general, if we have some relation "(1) . . . (2)," where in the blank we fill in something like "is taller than," or "dislikes," or "electrocuted," and we have some statement of the form "*b* . . . *c*," where *b* and *c* denote individuals, then in deciding in what order to write *b* and *c* after the relation symbol, we match up the statement with the relation and see which individual constant corresponds to which number. The individual constant corresponding to (1) is written first, and the individual constant corresponding to (2) is written second.

The relation "is taller than" is a *two-place relation*—it relates two things to one another. But some relations relate more than two things. The relation "is between" relates three things to one another; for example, we may say that John is between Bob and William. So "is between" is a three-place relation. In symbolizing statements involving this relation we might let B^3 stand for "(1) is between (2) and (3)."

Then letting j stand for "John," b for "Bob," and m for "William," we can symbolize the form of "John is between Bob and William" as B^3jbm. Had we instead let B^3 stand for "(3) is between (1) and (2)," then the statement would be symbolized as B^3bmj. The superscript on a relation symbol indicates how many places the relation has; T^2 stands for a two-place relation, and B^3 stands for a three-place relation. The superscript tells us how many individual constants must be written after the relation symbol in order to symbolize a statement.

In order to simplify our terminology, it is customary to call predicate letters and sentential letters relation symbols too. We can think of a predicate as a one-place relation, and a statement as a zero-place relation.

III.2 EXERCISES

A. Letting T^2 be "(1) is taller than (2)" and B^3 be "(1) is between (2) and (3)," and letting j stand for "John," b for "Bob," and m for "William," symbolize the forms of the following statements:

1. If Bob is taller than John, then it is not the case that John is taller than Bob.
2. If John is taller than Bob, and Bob is taller than William, then John is taller than William.
3. It is not the case that John is taller than himself.
4. If John is between Bob and William, then John is between William and Bob.
5. If John is between Bob and William, then neither is Bob between John and William nor is William between John and Bob.

B. Letting B^3 be "(3) is between (1) and (2)," symbolize the fourth and fifth statements of the preceding exercise.

III.3 QUANTIFIERS AND VARIABLES

We are still not able to symbolize the forms of statements like "All apples are red." Now we must introduce the concepts of a variable and a quantifier. Consider the statement "Given any two things, either the first is taller than the second, or the second is taller than the first." In order to symbolize the form of this statement, we must introduce some symbolic device that will work like the expressions "the first" and "the second" to allow us to talk about two unspecified objects. To do this we borrow the notion of a variable from algebra. In algebra we write things like $x + y = y + x$. This means "Given any two numbers, the first added to the second is equal to the second added to the first."[2] In other words, we use the variables x and y instead of the expressions "the first" and "the second." Using variables in this way, we could paraphrase the above statement as "Given any two objects, x and y, either x is taller than y, or y is taller than x." Similarly, we can paraphrase a statement like "Given any object, there is an object taller than it" as "Given any object x, there is an object y such that y is taller than x." In other words, variables act much like pronouns. They allow us to refer to the same object several times and to keep it separate from other objects we may be talking about.

Let us say precisely what symbols we will use as variables. A *variable* will be any lower case letter of the alphabet from t through z, with or without numerical subscripts; for example, x, z, t_2, and w_{137} are variables.

We can paraphrase statements in which "every," "all," "any," "some," and "there is" occur, using variables. "Given any two objects, if the first is taller than the second, then it is not the case that the second is taller than the first" can be paraphrased as "Given any two objects, x and y, if x is taller than y, then it is not the case that y is taller than x." If we now write variables after relation symbols just like we wrote individual constants after them, we can partially symbolize this statement as "Given any two objects, x and y, if T^2xy then it is not the case that T^2yx," or "Given any two objects, x and y, $(T^2xy \supset \sim T^2yx)$." Simple statements like "Everything is red" can also be paraphrased in this way: "Given anything, x, x is red."

Now consider how we might symbolize the form of the statement "All apples are red." First we paraphrase it to get it into the form "Given anything, x, . . ." What can we put in the blank? "All apples are red" means "Everything that is an apple is red," or "Given anything, if it is an apple then it is red." So this can be paraphrased as "Given anything, x, if x is an apple then x is red." Letting A^1 be "is an apple" and R^1 be "is red," we can partially symbolize this as "Given anything, x, $(A^1x \supset R^1x)$." Now all that remains is to introduce some way to symbolize the phrase "Given anything, x." We will

[2] Notice that when we write "Given any two numbers," we do not mean "Given any two *different* numbers." The two numbers may be equal; for example x and y in $x + y = y + x$ may both refer to the same number.

do this simply by enclosing x in parentheses: "(x)." Then, "Given anything, x, $(A^1x \supset R^1x)$" can be symbolized as $(x)(A^1x \supset R^1x)$; "(x)" is called the *universal quantifier with respect to* x. We can also talk about "(y)," the universal quantifier with respect to y, and so on. Note that it makes no difference which particular variable we use in symbolizing something. We can symbolize "All apples are red" as either $(x)(A^1x \supset R^1x)$ or $(y)(A^1y \supset R^1y)$. The reason for having more than one variable at our disposal will become apparent when we discuss formulas that contain more than one quantifier.

Many statements, such as "All apples are red," will have the form "All A are B." It is worthwhile to see how these statements are symbolized in general. To say that all A are B is the same as to say that, "Given anything x, if x is A then x is B." So "All A are B" can be partially symbolized as (x)(if x is A then x is B). Letting A^1 represent "is A" and B^1 represent "is B," we can further symbolize this as $(x)(A^1x \supset B^1x)$. In general, "All A are B" can be symbolized as $(x)(A^1x \supset B^1x)$.

In symbolizing the forms of some statements we will have to use more than one quantifier; for example, "Given any two things, if the first is taller than the second, then it is false that the second is taller than the first," we can first partially symbolize as "Given any two things, x and y, $(T^2xy \supset \sim T^2yx)$." This in turn can be paraphrased as "Given anything x, given anything y, $(T^2xy \supset \sim T^2yx)$," which can then be symbolized as $(x)(y)(T^2xy \supset \sim T^2yx)$.

The universal quantifier corresponds to the expressions "all" and "every." It is convenient to have another kind of quantifier corresponding to the expressions "some" and "there exists." Consider the statement "Some apples are green." This can be paraphrased as "There is something, x, which is a green apple." To say that something is a green apple is just to say that it is both green and an apple. So "Some apples are green" can be paraphrased as "There is something, x, such that $(G^1x \& A^1x)$." Next we want to introduce a symbol for the phrase "There is something, x, such that." The symbol "$(\exists x)$" will be used. This is called *the existential quantifier with respect to* x. Then "Some apples are green" can be symbolized as $(\exists x)(G^1x \& A^1x)$. In general, a statement of the form "Some A are B" or "There exists an A that is a B" can be symbolized as $(\exists x)(A^1x \& B^1x)$.

It should be noted that in symbolizing a statement of the form "Some A are B" we use a conjunction, whereas in symbolizing a statement of the form "All A are B" we use a conditional. There seems to be a strong temptation to mix these up, and either symbolize "All A are B" as $(x)(A^1x \& B^1x)$ or "Some A are B" as $(\exists x)(A^1x \supset B^1x)$. These are wrong for the following reasons: First consider $(x)(A^1x \& B^1x)$. This says that everything is *both* an A and a B. But that is not what we mean when we say that *All A are B*; we do not mean to imply that everything is an A. Thus we must write $(x)(A^1x \supset B^1x)$ rather than $(x)(A^1x \& B^1x)$. Next consider $(\exists x)(A^1x \supset B^1x)$. Recalling that $(A^1x \supset B^1x)$ is tautologically equivalent to $(\sim A^1x \vee B^1x)$, this formula is equivalent to $(\exists x)(\sim A^1x \vee B^1x)$. But this merely says that there is something which either is not an A or is a B. This will be true if there is something which is not an A, regardless of whether there is anything that is a B. For example, if G^1 is "is green" and A^1 is "is an apple," then $(\exists x)(\sim A^1x \vee G^1x)$ would be true just as long as there is something that is not an apple. It does not require that any apples be green. Consequently, we must write $(\exists x)(A^1x \& G^1x)$ rather than $(\exists x)(A^1x \supset G^1x)$.

A formula of the form $(x)(\ldots)$ is called a *universal generalization* (with respect to x), and a formula of the form $(\exists x)(\ldots)$ is called an *existential generalization* (with respect to x). Formulas that can be constructed using universal and existential quantifiers, sentential letters, predicate letters, relation symbols, individual constants, variables, and sentential connectives, are called *formulas of the predicate calculus*.

Now let us consider some examples of symbolizing the forms of statements using formulas of the predicate calculus. Let P^1 stand for the predicate "is a positive integer," let E^1 stand for "is even," let O^1 stand for "is odd," let L^2 stand for "(1) is less than (2)," and let I^2 stand for "(1) is equal to (2)." Let a_1 stand for the numeral "1," a_2 for the numeral "2," a_3 for the numeral "3," and so on.

First consider some statements not involving quantifiers. The statement "1 is less than 2" will be symbolized as $L^2a_1a_2$. The statement "1 is less than or equal to 2" can be paraphrased as "1 is less than 2, or 1 is equal to 2," which can then be symbolized as $(L^2a_1a_2 \vee I^2a_1a_2)$. The statement "7 is greater than 5," can first be paraphrased as "5 is less than 7," and then symbolized as $L^2a_5a_7$.

Now consider some statements involving quantifiers. The statement "Every positive integer is either odd or even" can be paraphrased as "Given anything x, if x is a positive integer then either x is odd or x is even." This can then be partially symbolized as (x)[if P^1x then either O^1x or E^1x], and then finished as $(x)[P^1x \supset (O^1x \vee E^1x)]$.

The statement "There is a positive integer greater than every positive integer" can be paraphrased as "There is something x, such that x is a positive integer, and x is greater than every positive integer." This in turn can be partially symbolized as $(\exists x)[P^1x$ and x is greater than every positive integer]. This can be paraphrased as $(\exists x)[P^1x$ and given anything y, if y is a positive integer then x is greater than $y]$. This can be partially symbolized as $(\exists x)[P^1x$ and $(y)($if P^1y then $L^2yx)]$, and then finished as $(\exists x)[P^1x \;\&\; (y)(P^1y \supset L^2yx)]$.

The symbolization of the last statement illustrates the general pattern that should be followed in complex cases. The process of symbolizing is done in two steps: first we introduce quantifiers and variables; second we symbolize the rest as in the propositional calculus, beginning with the smallest parts and working outwards. It is illuminating to diagram this process as follows:

There is a positive integer greater than every positive integer.

$(\exists x)[x$ is a positive integer and x is greater than every positive integer]

$(\exists x)[x$ is a positive integer and $(y)($if y is a positive integer then x is greater than $y)]$

$(\exists x)[\quad P^1x \quad \& \quad (y)($if P^1y then $L^2yx)]$

$$(\exists x)[P^1x \;\&\; (y)(P^1y \supset L^2yx)]$$

In the first step of symbolizing we begin with the outermost (leftmost) quantifier and then work inwards, introducing new quantifiers and variables as we go; we begin with the largest parts and work down to the smallest parts. Then in the second step, after we have introduced all of the quantifiers in the formula or in a part of the formula, we symbolize as in the propositional calculus, beginning with the smallest parts and working outwards to the larger parts. It should be noticed that "(y)" comes in the middle of the formula—just where it comes in paraphrasing the original sentences. It must be emphasized that the quantifiers are not automatically placed at the front of the formula. It is a very common mistake always to try to put them there.

Next consider the statement "Given any positive integer, there is a greater positive integer." First, introducing quantifiers and variables, we proceed as follows, from the outside working in:

Given any positive integer, there is a greater positive integer.

(x)[if P^1x then $\underbrace{\text{there is a positive integer greater than } x}$]

(x)[if P^1x then $(\exists y)(P^1y$ and $L^2xy)$]

$(x)[P^1x \supset (\exists y)(P^1y \,\&\, L^2xy)]$

The statement "Given any odd positive integer, there is a greater even positive integer" can be symbolized as follows:

Given any odd positive integer, there is a greater even positive integer.
(x)[if x is an odd positive integer, then there is an even positive integer greater than x]
(x)[if $(P^1x \,\&\, O^1x)$ then there is an even positive integer greater than x]
(x)[if $(P^1x \,\&\, O^1x)$ then $(\exists y)(y$ is an even positive integer, and y is greater than $x)$]
$(x)[(P^1x \,\&\, O^1x) \supset (\exists y)([P^1y \,\&\, E^1y] \,\&\, L^2xy)]$

Now consider the statement "No positive integer is less than itself." If we are to symbolize this statement in the predicate calculus, "No" must be paraphrased in terms of our quantifiers. This can be done as follows:

No positive integer is less than itself.
It is not the case that there is a positive integer less than itself.
$\sim(\exists x)(x$ is a positive integer and x is less than itself$)$
$\sim(\exists x)(P^1x \,\&\, L^2xx)$

In general, "No A is B" can be paraphrased either as "It is not the case that there is an A that is B" or as "Given any A, it is not a B." We used the first paraphrase above, giving us something of the form $\sim(\exists x)(A^1x \,\&\, B^1x)$. Had we used the second paraphrase we would have had something of the form $(x)(A^1x \supset \sim B^1x)$, or, in the above case, $(x)(P^1x \supset \sim L^2xx)$. It will be shown later that the formulas $\sim(\exists x)(A^1x \,\&\, B^1x)$ and $(x)(A^1x \supset \sim B^1x)$ are equivalent, so it makes no difference which way we symbolize a statement involving "No."

It should be reemphasized that, in symbolizing statement forms in the predicate calculus, one should begin with the outermost quantifiers and work inwards when introducing quantifiers and variables. Then those parts that do not involve quantifiers can be symbolized just as in the propositional calculus.

When we symbolize the form of a statement and introduce a quantifier together with several occurrences of the corresponding variable, we say that the quantifier *binds* those occurrences of the variable. Therefore, the quantifier "(x)" binds both occurrences of x in $(x)(A^1x \supset B^1x)$.[3] Similarly, the quantifier "(x)" binds all occurrences of x and the quantifier "$(\exists y)$" binds all occurrences of y in the formula $(x)(A^1x \supset (\exists y)R^2xy)$. It should be noticed however that different occurrences of the same variable can be bound by different quantifiers. In symbolizing the statement "Either all apples are red or all apples are green" we get the formula $[(x)(A^1x \supset R^1x) \vee (x)(A^1x \supset G^1x)]$. Here the first and second occurrences of x are bound by the first quantifier, and the third and fourth occurrences are bound by the second quantifier. To take a more complex case,

$$(x)[(A^1x \supset B^1x) \ \& \ (\exists x)A^1x]$$

the universal quantifier binds the first two occurrences of x, and the existential quantifier binds the third occurrence of x. An occurrence of a variable is always bound by the innermost quantifier that can bind it. Thus in the above example the third occurrence of x is bound by the existential quantifier rather than the universal quantifier. If we were to omit the existential quantifier, then all three occurrences of x would be bound by the universal quantifier. In general, if we have a formula of the form $(x)(\ldots x \ldots)$ or $(\exists x)(\ldots x \ldots)$, the quantifier binds all occurrences of x within the formula that are not already bound by another quantifier. Notice also that a quantifier only binds occurrences of the variable that appears in the quantifier. So "(x)" can only bind occurrences of x, and "(y)" can only bind occurrences of y.

That formula that immediately follows a quantifier, in which any occurrences of the variable of that quantifier not previously bound are bound by the quantifier, is called the *scope* of the quantifier. So, in the formula $[(x)(A^1x \supset R^1x) \vee (x)(A^1x \supset G^1x)]$, $(A^1x \supset R^1x)$ is the scope of the first quantifier, and $(A^1x \supset G^1x)$ is the scope of the second quantifier. In the formula $(x)(A^1x \supset (\exists y)R^2xy)$, $(A^1x \supset (\exists y)R^2xy)$ is the scope of "(x)," and R^2xy is the scope of "$(\exists y)$." An occurrence of a variable is bound by a quantifier if, and only if, the quantifier contains that variable, and the occurrence in question is within the scope of that quantifier and not already bound by another quantifier.

[3] The x in the quantifier is not considered an occurrence of the variable.

Notice that in a formula we get by symbolizing the form of a statement, each occurrence of each variable must be bound by some quantifier. This is because we only introduce variables along with quantifiers that bind them. Variables only come in when we symbolize "all," "each," "every," "some," and so on.

Thus far we have been patiently writing superscripts on all of our relation symbols, but from now on we will generally omit them. There is no danger of confusion in doing so, because in any given formula we can always figure out what the superscript on a relation symbol should be just by counting the number of variables and individual constants that follow it; for example, in the formula $(x)(Hx \equiv Rxa)$, we know that "H" should have the superscript "1" and R should have the superscript "2."

Existential and universal quantifiers can be used to symbolize a good many English phrases. "All," "each," and "every" can all be symbolized using the universal quantifier, and "some," "there is," and "there exists" can be symbolized using the existential quantifier. "No" and "none" can be symbolized in different ways using either the existential or universal quantifier. However, some care must be exercised here because statements should not always be symbolized in the way it may at first appear they should be. The statement "If someone is dead, there is a murderer in the house" is correctly symbolized as $[(\exists x)(Px \ \& \ Dx) \supset (\exists y)(My \ \& \ Hy)]$, where P stands for "is a person," D for "is dead," M for "is a murderer," and H for "is in the house." Similarly, there is a strong temptation to symbolize "If someone is dead, Jones killed him" as $[(\exists x)(Px \ \& \ Dx) \supset Kjx]$. However, this cannot be right, because the x in the consequent is not bound by any quantifier. If one stops and thinks what this statement *means*, it can be seen that, in spite of the "someone," it is really a universal statement meaning "Given anyone, if he is dead then Jones killed him." It is properly symbolized using a universal quantifier: $(x)[(Px \ \& \ Dx) \supset Kjx]$.

III.3 EXERCISES

A. Letting P^1 stand for "was a president of the United States," B^2 for "(1) was president before (2)," D^2 for "(1) died before (2)," A^2 for "(1) was born before (2)," and P^1_2 for "is a person," and letting g

stand for "Washington," l for "Lincoln," and j for "Jefferson," symbolize the forms of the following statements:

1. Both Washington and Lincoln were presidents of the United States.
2. Washington was president before Lincoln.
3. There was no president before Washington.
4. Washington was president before anyone who was president after Lincoln.
5. Someone was president before Lincoln and after Jefferson.
6. Everyone who was president before Jefferson was president before Lincoln.
7. Given any two presidents, if the first was born before the second, then the first died before the second.
8. Some presidents were born before others, and some died after others.
9. There is at least one president who was born before another president and died after him.
10. There were no two presidents between Jefferson and Lincoln such that one was born before the other but died after him.
11. There is a president who was president both before and after another president.
12. Given any three past presidents, if the first died before the second, and the second died before the third, then the first died before the third, but it is not the case that given any three presidents, if the first was born before the second, and the second was president before the third, then the third died after the second.
13. Given any two presidents, if the first died before Jefferson, and Lincoln died before the second, then if the second was born before the first, then the second was president before the first and died after Washington but before Jefferson.
14. No president was president both before and after two presidents who in turn were each president both before and after each other.

B. In the following formulas, indicate the scope of each quantifier, and which occurrences of variables are bound by each quantifier:

1. $(x)(Ax \supset Bx)$
2. $(x)(Axb \supset (\exists y)Hxy)$
3. $[Rab \vee (x)(\exists y)Hxy]$
4. $(x)[Hxa \supset \{(Fax \equiv Hbx) \supset \sim(\exists x)Hxx\}]$
5. $(x)(y)(z)Hxyz$
6. $[(x)(Ax \supset Bx) \vee \sim(\exists x)(Ax \;\&\; \sim Bx)]$
7. $(x)\{Hx \supset (x)[Hx \supset (x)(Hx \supset (x)Hx)]\}$

C. Letting D stand for "is a dog," I for "is an Indian," T for "is dead," A for "is a day," S for "is a stone," L for "is an angel," R for "is a time when it rains," B for "is rolling," P for "is a time when it pours," F for "is a fool," M for "is moss," O for "is good," H for "(1) has (2)," U for "(1) rushes into (2)," G for "(1) gathers (2)," and E for "(1) fears to tread in (2)," the following formulas symbolize familiar sayings. Express these sayings in English.

1. $(x)(Rx \supset Px)$
2. $(x)[(Ox \& Ix) \supset (Tx \& Ix)]$
3. $(x)[Dx \supset (\exists y)(Ay \& Hxy)]$
4. $(x)[(Sx \& Bx) \supset \sim(\exists y)(My \& Gxy)]$
5. $(x)(y)\{[(\exists z)(Lz \& Ezy) \& Uxy] \supset Fx\}$

III.4 SYMBOLIZING THE FORMS OF RELATIONS

Up to this point we have only been concerned with symbolizing the forms of statements, but the same thing can be done with relations. Some relations can be thought of as contructed out of other simpler relations; for example, the relation "Either (1) loves (2), or (2) loves (1)" is constructed out of the simpler relation "(1) loves (2)." If we let the relation symbol L stand for the simpler relation "(1) loves (2)," we can symbolize the form of the complex relation by using this relation symbol and writing $[L(1)(2) \vee L(2)(1)]$. We can also go a step further and put variables in for the numbers in parentheses. Then we get the formula $(Lxy \vee Lyx)$, which gives us the logical form of the complex relation.

Let us look at a more complicated example. Consider the relation "If (1) strikes (2) in the presence of (3), then (3) will think poorly of (1), but (2) will be embarrassed and will think poorly of himself." Letting S stand for "(1) strikes (2) in the presence of (3)," letting T stand for "(1) thinks poorly of (2)," and letting E stand for "is embarrassed," we can symbolize the complex relation as

If $S(1)(2)(3)$, then $T(3)(1)$, but $E(2)$ and $T(2)(2)$
$[S(1)(2)(3) \supset (T(3)(1) \& E(2) \& T(2)(2))]$

If we then replace the numbers by variables we get

$$[Sxyz \supset (Tzx \& Ey \& Tyy)]$$

Notice that we pick a variable to substitute for each number in parentheses each time it occurs; for example, we put x in place of "(1)" each time it occurs, and y in place of "(2)." It makes no difference which variable we choose. We could have symbolized the above relation equally well as $[Szyx \supset (Txz \,\&\, Ey \,\&\, Tyy)]$.

Sometimes, in symbolizing the form of a relation, we will have to use quantifiers. The relation "(1) is the father of the mother of (2)" can be paragraphed as "There is someone, x, such that (1) is the father of x and x is the mother of (2)." Letting P stand for "is a person," F stand for "(1) is the father of (2)," and M stand for "(1) is the mother (2)," we can symbolize this as $(\exists x)(Px \,\&\, F(1)x \,\&\, Mx(2))$. We can then put y in for "(1)" and z in for "(2)" to get

$$(\exists x)(Px \,\&\, Fyx \,\&\, Mxz)$$

Notice that the variables we substitute for "(1)" and "(2)" must be different from the variable we use with the quantifier. Had we used x for "(1)" and y for "(2)," getting the formula $(\exists x)(Px \,\&\, Fxx \,\&\, Mxy)$, we would instead have had a formula meaning "There is someone who is his own father, and he is the mother of (1)." Each time we introduce a variable to serve a new purpose in symbolizing the form of a relation, it must be distinct from any we have used before.

The formulas we get from symbolizing the forms of relations are also called formulas of the predicate calculus, but they are different in an important respect from the formulas we get from symbolizing the forms of statements. In the formulas that result from symbolizing the forms of statements, each occurrence of each variable must be bound by some quantifier. This is because variables were only introduced in symbolizing the expressions "all," "each," and so on; the quantifiers that bind them were introduced at the same time. But in symbolizing the forms of relations, the variables that we put in place of the numbers in parentheses are not bound by any quantifiers. If an occurrence of a variable in a formula is not bound by any quantifier, it is called a *free* occurrence; for example, the occurrence of z in $(x)(\exists y)(Fxy \equiv Hxyz)$ is free. Hence, the difference between the formulas that result from symbolizing the forms of statements and those that result from symbolizing the forms of relations is that the former never have free occurrences of variables, and the latter always have free occurrences of variables.[4]

[4] Some authors prefer to call predicate letters and relation symbols variables too. In that case, the symbols that are here called variables must be termed *individual variables* to set them apart from the others. Then the above point can be made by saying that a formula containing free occurrences of individual variables symbolizes the form of a relation, and one containing no free individual variables symbolizes the form of a statement.

A formula of the predicate calculus that does not have any free occurrences of variables is called a *sentence* of the predicate calculus. So we always get sentences when we symbolize the forms of statements.

III.4 EXERCISES

A. Symbolize the forms of the following relations, letting M stand for "(1) is married to (2)," W stand for "(1) is a wife of (2)," F stand for "(1) is the father of (2)," P stand for "(1) is a parent of (2)," B stand for "(1) is a brother of (2)," S stand for "(1) is a sister of (2)," D stand for "(1) is a daughter of (2)," and P_2 stand for "is a person."

1. (1) is a grandfather of (2).
2. (1) is the paternal grandfather of (2).
3. (1) is a sister-in-law of (2).
4. (1) is the paternal grandfather of (2)'s maternal grandmother.
5. (1) is a cousin of (2).
6. (1) is a niece of (2).
7. (1) is a nephew of (2).
8. (1) is an uncle of (2).
9. (1) is a cousin of a nephew of an uncle of (2).
10. A cousin of a sibling of (1) is a nephew of (2) and a grandmother of (3).

B. For each of the following formulas, tell which (if any) occurrences of variables are free:

1. Bxa
2. $(x)(Fxa \supset Bxy)$
3. $[(\exists x)(Gx \,\&\, Hyz) \vee Hxx]$
4. $(x)[Hxa \equiv (\exists y)(\exists z)(Hyx \supset {\sim}Hzz)]$
5. $(x)(Hx_1 \supset Bxx)$
6. $[(x)(\exists y)(Bxyz \vee {\sim}Bzyx) \vee {\sim}(Bxyz \vee \supset Bzyx)]$

III.5 UNIVERSES OF DISCOURSE

It is customary to introduce a certain simplification into the process of symbolizing statement forms in the predicate calculus. This is the introduction of *universes of discourse*. What we do is limit the class of individuals to which the quantifiers refer. For instance, in the statements in the last section involving positive integers, we might have let "(x)" mean "Given any positive integer x" rather than "Given anything x," and let "$(\exists x)$" mean "There is a positive integer x, such that" rather than "There is something x, such that." This would have made things somewhat simpler. Consider the statement "Given any positive integer, there is a greater positive integer." Using this simplification this could be symbolized as follows:

Given any positive integer, there is a greater positive integer.
Given any positive integer x, there is a positive integer greater than x.
(x)(there is a positive integer y such that y is greater than x)
$(x)(\exists y)Lxy$

Compare this with our original result, which was

$$(x)[Px \supset (\exists y)(Py \,\&\, Lxy)]$$

To take another example, consider the statement "Someone is loved by everyone." Letting P stand for the predicate "is a person," and L stand for the relation "(1) loves (2)," this could be symbolized as follows:

Someone is loved by everyone.
There is something, x, such that x is a person, and x is loved by everyone.
$(\exists x)$(x is a person, and x is loved by everyone)
$(\exists x)$(Px and given anything y, if y is a person then y loves x)
$(\exists x)(Px \,\&\, (y)(Py \supset Lyx))$

However, if we let the quantifiers "(x)" and "$(\exists y)$" mean "Given anyone x" and "There is someone y such that" respectively, we could symbolize this as:

Someone is loved by everyone.
There is someone, x, such that x is loved by everyone.
$(\exists x)$(x is loved by everyone)
$(\exists x)$(given anyone, y, y loves x)
$(\exists x)(y)Lyx$

When we restrict the meaning of our quantifiers like this we say that we are restricting our universe of discourse—we are only talking about a certain restricted class of things. When this is done we must specify not only what predicates and relations our predicate letters and relation symbols stand for, and what individuals our individual constants denote, but also what the universe of discourse is. We can specify any universe of discourse we like. Given some class C of things as a universe of discourse, our quantifiers "(x)" and "$(\exists x)$" then mean "Given anything x which is in the class C" and "There is something x

in the class C such that" respectively. When using a restricted universe of discourse, we say our quantifiers *range over* the class C.

Now consider our universe of discourse to be the class of automobiles, let F be the predicate "is a Ferrari," M be the predicate "is a Maseratti," and let Q be the relation "(1) is faster than (2)." Now we can symbolize the following statements:

(1) There is an automobile that is faster than any Maseratti.
There is an automobile x such that x is faster than any Maseratti.
$(\exists x)$(x is faster than any Maseratti)
$(\exists x)$(given any automobile y, if y is a Maseratti then x is faster than y)
$(\exists x)(y)$ (if My then Qxy)
$(\exists x)(y)(My \supset Qxy)$

(2) There is a Ferarri such that no car is faster than it.
There is an x such that x is a Ferrari, and no car is faster than x.
$(\exists x)$[x is a Ferrari, and no car is faster than x]
$(\exists x)$[Fx and $\sim$(there is an automobile faster than x)]
$(\exists x)$[Fx & $\sim$(there is an automobile y such that y is faster than x)]
$(\exists x)[Fx \,\&\, \sim(\exists y)Qyx]$

(3) No car that is not a Maseratti is faster than any Maseratti.
$\sim$[There is a car that is not a Maseratti and is faster than some Maseratti]
$\sim$[There is an automobile x such that x is not a Maseratti, and there is a Maseratti such that x is faster than it]
$\sim(\exists x)$[$\sim Mx$ & there is a Maseratti such that x is faster than it]
$\sim(\exists x)$[$\sim Mx$ & there is a y such that My and Qxy]
$\sim(\exists x)[\sim Mx \,\&\, (\exists y)(My \,\&\, Qxy)]$

When we do not specify a universe of discourse, it is taken for granted that the universe is the class of everything. Then our quantifiers "(x)" and "$(\exists x)$" read simply "Given anything x" and "There is something x such that." But often, by specifying a restricted universe of discourse, we can make the task of symbolizing a group of statements much simpler.

III.5 EXERCISES

A. Taking the class of human beings as your universe of discourse, and letting M stand for "(1) is married to (2)," W stand for "(1) is a wife of (2)," F stand for "(1) is the father of (2)," P stand for "(1) is a parent of (2)," B stand for "(1) is a brother of (2)," S stand for "(1) is a sister of (2)," a stand for "Arthur," and b stand for "Bartholomew," symbolize the forms of the following statements:

1. It is false that anyone is both the father of Arthur and the mother of Bartholomew.
2. Arthur has a mother who is married to his father.
3. Arthur is the paternal grandfather of Bartholomew.
4. Arthur is a sibling of someone.
5. Everyone who is married to anyone is the offspring of someone.
6. One of Arthur's grandfathers is a brother of one of Bartholomew's grandmothers.
7. Anyone who is married to someone's brother is someone's sister-in-law.
8. No one is the father of a brother of his mother.

B. Symbolize the forms of (1), (4), and (7) without using a restricted universe of discourse.

III.6 FORMULAS OF THE PREDICATE CALCULUS

So far we have been content to deal with rather rough and ready notions of "formula of the predicate calculus" and "sentence of the predicate calculus." Now precise rules for constructing formulas and sentences will be formulated on the analogy of the rules given in Chapter II for constructing formulas of the propositional calculus.

We begin by saying that any capital letter, with or without numerical subscripts and superscripts, is a *relation symbol*. The relation symbols include the sentential letters and predicate letters. If a relation symbol has no superscript, then it is a sentential letter; if it has the superscript "1" it is a predicate letter. A relation symbol having some numerical superscript, say n, is an *n-place relation symbol*; for example, R_7^2 is a two-place relation symbol.

Any lower case letter, with or without numerical subscripts, is an *individual symbol*. The individual symbols will be divided into two classes—the individual constants, and the variables. The lower case letters from a through s, with or without numerical subscripts, are *individual constants*; and lower case letters from t through z, with or without numerical subscripts, are *variables*.

The notion of an atomic formula must next be defined. Once again, a sentential letter written by itself is an atomic formula; and expressions like F^2xa or $G^4yza_1b_7$ out of which we construct complex formulas are atomic formulas. So our precise definition is as follows: *Atomic formulas* are those expressions consisting either of a sentential letter written by itself, or a relation symbol having a numerical superscript followed immediately by a list of that number of individual symbols. So, for example, the following are all atomic formulas:

$$P,\ Q,\ A_{13},\ F^1x,\ F^3xxx,\ G^3xyz,\ A^2_{37}ab,\ B^4xaxa$$

Precise rules for constructing other formulas from atomic formulas can be formulated as follows:

1. An atomic formula is a formula.
2. If P is a formula, then $\sim P$ is a formula.
3. If P and Q are formulas, then $(P \,\&\, Q)$, $(P \vee Q)$, $(P \supset Q)$, and $(P \equiv Q)$ are formulas.
4. If P is a formula, and x is any variable, then $(x)P$ and $(\exists x)P$ are formulas.

The notions of free and bound occurrences of variables can also be made more precise. In a formula of the form $(x)P$ or $(\exists x)P$ constructed by Rule 4, P is called the *scope* of the quantifier. It should be observed that by requiring that P be a formula we ensure that in any given formula, there can only be one scope for each quantifier. In $((\exists x)Fx \vee P)$ for example, the scope of the quantifier must be Fx. It cannot be $Fx \vee P$ because this is not a formula (it lacks parentheses). An occurrence of a variable in a formula is *bound* if, and only if, it is within the scope of a quantifier containing that variable. An occurrence of a variable is *free* if, and only if, it is not bound. A formula that contains no free occurrences of variables is a *sentence*. An occurrence of a variable is bound by a particular quantifier if, and only if, the variable in the quantifier is the same as the variable in question, and the quantifier is the innermost quantifier whose scope contains the occurrence in question.

A peculiarity of our definition is that we can append a quantifier to a formula using Rule 4 even if that formula contains no free occurrences of x. A quantifier like this is said to be a *vacuous* quantifier. We could have framed Rule 4 in such a way as to preclude vacuous quantifiers, but it would then have become excessively complicated. A formula with a vacuous quantifier will be taken to mean the same thing as the corresponding formula without the quantifier; thus the formula $(x)((\exists y)F^1y \supset (\exists z)G^1z)$ means the same thing as

$$((\exists y)F^1y \supset (\exists z)G^1z)$$

A vacuous quantifier leaves the meaning of a formula unchanged.

Now look at some sample constructions of formulas using these rules. Consider the formula $(x)((P \mathbin{\&} F^1x) \supset (\exists y)G^2xy)$. It can be constructed using the above four rules. We can diagram the construction as follows.

$P \quad F^1x$	G^2xy	*Rule* (1)
$(P \mathbin{\&} F^1x)$		*Rule* (3)
	$(\exists y)G^2xy$	*Rule* (4)
$((P \mathbin{\&} F^1x) \supset (\exists y)G^2xy)$		*Rule* (3)
$(x)((P \mathbin{\&} F^1x) \supset (\exists y)G^2xy)$		*Rule* (4)

Next let us construct the formula $(x)(F^1x \mathbin{\&} (\exists x)H^2xx)$:

F^1x	H^2xx	*Rule* (1)
	$(\exists x)H^2xx$	*Rule* (4)
$(F^1x \mathbin{\&} (\exists x)H^2xx)$		*Rule* (3)
$(x)(F^1x \mathbin{\&} (\exists x)H^2xx)$		*Rule* (4)

For convenience, we will adopt the same conventions we had in Chapter II concerning the use of brackets in place of parentheses and the omission of parentheses inside a long conjunction or disjunction. We will also omit the superscripts on relation symbols. However, the expressions we get using these conventions are *not*, strictly speaking, formulas; they are only abbreviations for formulas.

It should be noticed that all formulas of the propositional calculus are also sentences of the predicate calculus. They can all be constructed using Rules 1 through 4.

Frequently we will want to talk about formulas that can be obtained from other formulas by substituting one individual symbol throughout a formula for every occurrence of another individual symbol. For example, the formula $(Fa \mathbin{\&} (x)Gxa)$ can be obtained from the formula $(Fy \mathbin{\&} (x)Gxy)$ by substituting a for every free occurrence of y. When we are substituting for a variable, it is only the free occurrences of the variable for which we substitute. Thus, we can substitute b for every free occurrence of x in $[(Hx \vee Ga) \mathbin{\&} (x)Hx]$ to obtain

$$[(Hb \vee Ga) \mathbin{\&} (x)Hx]$$

We only substitute for the free occurrences of the variables because a formula containing free occurrences of variables symbolizes the form of a complex relation; if we substitute individual constants for the free occurrences of the variables, we construct a sentence which says that the objects denoted by those individual constants stand in that relation to one another. This is the intended significance of the substitution of individual constants for variables.

A special notation will be introduced to stand for the result of substituting one individual symbol for another in a formula. Given any two individual symbols, α and β, and any formula P, $Sb(\genfrac{}{}{0pt}{}{\alpha}{\beta})P$ is the formula that results from substituting α for every occurrence of β in P, if β is an individual constant, and it is the formula that results from substituting α for every free occurrence of β if β is a variable. So, we have the following:

$$Sb(\genfrac{}{}{0pt}{}{a}{x})[Ha \vee \sim Hx] \text{ is } [Ha \vee \sim Ha]$$
$$Sb(\genfrac{}{}{0pt}{}{x}{a})[Ha \vee \sim Hx] \text{ is } [Hx \vee \sim Hx]$$
$$Sb(\genfrac{}{}{0pt}{}{x}{y})[(y)(Fxy \supset \sim Gay) \vee (x)(z)(Fxz \supset \sim Fay)] \text{ is }$$
$$[(y)(Fxy \supset \sim Gay) \vee (x)(z)(Fxz \supset \sim Fax)]$$

This can be generalized somewhat. Sometimes we may want to make more than one substitution at one time. In general, if $\alpha_1, \ldots, \alpha_n$ is a list of individual symbols (all different), and $\beta_1, \ldots, \beta_n$ is another list of individual symbols (not necessarily all different), and P is any formula; then $Sb(\genfrac{}{}{0pt}{}{\beta_1,\ldots,\beta_n}{\alpha_1,\ldots,\alpha_n})P$ is the formula that results from simultaneously substituting β_1 for each occurrence of α_1 if α_1 is an individual constant, or for each free occurrence of α_1 if α_1 is a variable, β_2 for each occurrence of α_2 if α_2 is an individual constant, or for each free occurrence of α_2 if α_2 is a variable, and so on. For example

$$Sb(\genfrac{}{}{0pt}{}{a,z,x,x}{x,y,a,b})[((x)Hx \vee Fxay) \equiv Hb] \text{ is } [((x)Hx \vee Faxz) \equiv Hx]$$

III.6 EXERCISES

A. Which of the following are formulas of the predicate calculus? Which are sentences?

1. $(x)(y)(Fxy \supset Fyx)$
2. $(P \supset (\exists x)Gxy)$
3. $((x)(\exists y)(z)Gxy \;\&\; Fyz)$
4. $(\exists z)(x)Fxz \supset (x)(\exists y)Gxy$
5. $(\exists x) \sim (P \;\&\; (x)(Fx \supset Gx))$
6. $(x)(y)((Bxa \supset Bxy) \vee \sim(Hxyz \;\&\; Bzy))$
7. $(x)(Fx) \supset (\exists y)(Gxy \supset Fy)$
8. $((w)Pw \equiv \sim(\exists z)(Wp \;\&\; Pz))$
9. $(Wa \;\&\; \sim Wa)$
10. $((P \;\&\; Q) \supset R)$

B. Determine what formulas result from making the following substitutions:

1. $Sb\binom{a}{x}Fx$
2. $Sb\binom{a}{x}Fa$
3. $Sb\binom{a}{x}[Fx \supset (x)Gx]$
4. $Sb\binom{b,c,x}{x,a,b}[(x)(\exists y)Facyx \supset (Gxba \;\&\; \sim Fcbcc)]$
5. $Sb\binom{w,y,z}{w,c,a}[(\exists y)Fyw \;\&\; ((\exists y)Fya \supset (z)(\exists y)Fzya)]$
6. $Sb\binom{b,c,a}{a,b,c}(Fabc \vee \sim(Fbba \;\&\; Fcab))$
7. $Sb\binom{x,x,y,z,b,c,w}{y,z,x,c,w,b,a}\{[Fabwzd \equiv \sim Fbcdzw) \;\&\; \sim(Wwwbcwa \vee \sim P)] \supset \sim Bxzzy\}$

C*. Explain why we cannot simply define $Sb\binom{\beta_1,\ldots,\beta_n}{\alpha_1,\ldots,\alpha_n}P$ as $Sb\binom{\beta_1}{\alpha_1}Sb\binom{\beta_2}{\alpha_2}\ldots Sb\binom{\beta_n}{\alpha_n}P$.

III.7 THE RUDIMENTS OF SET THEORY

We want to analyze the truth conditions of statements whose forms can be symbolized in the predicate calculus, just as we did for similar statements in the propositional calculus. However, the situation is considerably more complex in the predicate calculus than it was in the propositional calculus. Before we can proceed, the student must have some acquaintance with set theory.

First then, what is a set? This is a question that we cannot really answer except by giving synonyms. The terms "set," "class," "collection," "group," and "ensemble" can all be used interchangeably. A set may be defined to be any collection of objects. For example, we can talk about the set of all integers between one and seventeen; this set consists of the integers from two through sixteen, inclusive. Or we may talk about the set of all desks in a room, or the set of all buildings on the campus, or the set of all past Presidents of the United States, and so forth.

A set is a set *of* objects. The objects in a set are called the *members* or *elements* of the set. Given any set, A, and any object, x, we abbreviate the statement "x is a member of A" by writing $x \in A$. Thus, for example, if A is the set of all past Presidents of the United States, and k stands for "John F. Kennedy," we can write $k \in A$. Or, if N is the set of integers between one and sixteen, then we can write $6 \in N$.

Perhaps the most important property of sets is that they are *extensional.* This means that the identity of a set is determined entirely by what objects it has as members—if two sets have exactly the same members, then they are the same set. For example, the set of all integers between one and four is identical to the set of the two integers two and three, because these sets have the same members. Or we can say that the set of featherless bipeds is the same as the set of things that are either human beings or plucked birds. Using the symbols of the predicate calculus, we can express what it means to say that sets are extensional as follows: Given any two sets, A and B,

$$[(x)(x \in A \equiv x \in B) \supset A = B]$$

Because the identity of a set is determined by its members, one convenient way to refer to a set having a small number of members is by listing its members. The customary way to do this is to enclose the list of members in braces, { and }. We can, for example, refer to the set of all integers between one and four by writing $\{2, 3\}$. It makes no difference in what order you list the members of the set. If the members are the same, the set is the same. So $\{2, 3\}$ and $\{3, 2\}$ are the same set, and $\{1, 2, 3\}$, $\{1, 3, 2\}$, $\{2, 1, 3\}$, $\{2, 3, 1\}$, $\{3, 1, 2\}$, and $\{3, 2, 1\}$ are the same set—the set whose members are the integers one, two, and three.

Sometimes a set will have just one member; for example, we can talk about the set of all integers between one and three, which is the set whose only member is two. This set can be referred to as $\{2\}$. A set having only one member is called the *unit set* of that object. So $\{2\}$ is the unit set of two.

Frequently a set will have too many members to make it practical to list all of them. A set might even have an infinite number of members, like the set of all integers. When this happens we cannot refer to the set by simply listing the members, but we can adapt the above notation to take care of this. If for some predicate A we want to talk about the set of all objects that are A (such as the set of all past presidents), we can simply write $\{x; x \text{ is } A\}$. This is read "the set of all objects, x, such that x is A." Using these symbols the set of all past presidents is $\{x;$ x is a past president$\}$, and the set of all integers is $\{x;\ x$ is an integer$\}$.

There is one set that deserves special mention because of its somewhat paradoxical-seeming nature. This is the set that has no members. This is called the *empty set*, and is designated by the symbol $\varnothing$. The empty set can be introduced in many different ways; for example, $\varnothing = \{x;\ x$ both is and is not an integer$\}$ because there is not anything that both is and is not an integer. Notice that there can only be one empty set—this follows from the extensionality of sets. If we had two sets neither of which had any members, then they would have the same members (that is, $(x)(x \in A \equiv x \in B)$ would be true), and so they would be the same set.

A set that is just the opposite of the empty set is the *universal set*. The universal set is the set containing everything. It contains all numbers, material objects, people, and so on. The universal set is denoted by U. We can define it as $U = \{x;\ x$ either is or is not an integer$\}$ because everything either is or is not an integer.

The order in which we list the members of a set makes no difference to the identity of the set. But sometimes we would like it to make a difference. So this brings us to a new sort of entity, which is called a *sequence*. A sequence is a list of objects having a definite fixed order; for example, we may talk about the sequence of integers from one to ten, listed in order from smallest to largest. This will be distinct from the sequence of integers from one to ten listed in order from largest to smallest.

Just as $\{$ and $\}$ are used to talk about sets, $\langle$ and $\rangle$ are used to talk about sequences. Thus, the sequence of integers from one to ten listed in order from smallest to largest will be the sequence $\langle 1, 2, 3, 4, 5, 6, 7, 8, 9, 10\rangle$, and the sequence of integers from one to ten listed in the reverse order will be $\langle 10, 9, 8, 7, 6, 5, 4, 3, 2, 1\rangle$. These are different sequences; unlike sets, sequences are not extensional. Whereas the identity of a set is determined solely by its members, the identity of a sequence is determined by the combination of its members *and* their order.

We will frequently want to talk about sequences of fixed length. A sequence of two objects will be called an *ordered pair*. Thus $\langle 1, 2\rangle$, $\langle$John Kennedy, F. Scott Fitzgerald$\rangle$, $\langle$Hiroshima, $\sqrt{2}\rangle$ are all ordered pairs. Similarly, a sequence of three objects is called an *ordered triple*, a sequence of four objects is called an *ordered quadruple*. Above a certain point we run out of names like this, so we use another kind of name.

An ordered pair can also be called an ordered two-tuple, and ordered triple can be called an ordered three-tuple, and so on. This terminology will work for a sequence of any length; a sequence of 137 objects can be called an ordered 137-tuple. In general, given any number n, we can talk about sequences of n objects and call them *ordered n-tuples.*

There is a close relationship between sets and relations. The simplest relations are one-place relations, or predicates. Corresponding to each predicate is the set of all objects to which that predicate can be truly ascribed. For example, corresponding to the predicate "is a bachelor" there is the set of all bachelors; that is $\{x; x \text{ is a bachelor}\}$. And corresponding to the predicate "is an integer" there is the set of all integers; that is, $\{x; x \text{ is an integer}\}$. In other words, given any predicate A, there corresponds to it $\{x; x \text{ is } A\}$. This set is called the *extension* of the predicate A. An object is a member of the extension of a predicate if, and only if, that predicate can be truly ascribed to that object.

Now let us look at two-place relations, such as "(1) is the brother of (2)." There is also a set corresponding to this, but now it is a set of ordered pairs; that is, corresponding to the relation "(1) is the brother of (2)" is the set of all ordered pairs which are such that the first element of the pair is the brother of the second element of the pair. Or corresponding to the relation "(1) is an integer less than (2), and (2) is an integer less than four," we have the set of all ordered pairs of integers such that the first integer in each pair is less than the second integer, and the second integer is less than four. This is just the set $\{\langle 1, 2\rangle, \langle 1, 3\rangle, \langle 2, 3\rangle\}$. In general, given a two-place relation "(1) . . . (2)" (where we fill something in for the blank), we can talk about the set of all ordered pairs which are such that the first member of each pair stands in this relation to the second member of that pair. This set can be referred to as "the set of all ordered pairs, $\langle x, y\rangle$, such that $x \ldots y$," which can be symbolized as $\{\langle x, y\rangle; x \ldots y\}$. Corresponding to the relation "(1) is the brother of (2)" is $\{\langle x, y\rangle; x \text{ is the brother of } y\}$. This set of ordered pairs is called the *extension* of the relation.

This concept of the extension of a relation can be generalized to relations of any number of places. In general, for any number, n, if we are given an n-place relation, A, the set of n-tuples $\{\langle x_1, x_2, \ldots, x_n\rangle; Ax_1x_2 \ldots x_n\}$ is the *extension* of the relation. The extension of the relation "(1) is an integer between (2) and (3), and (2) is greater than four and (3) is less than nine" is the set $\{\langle 6, 5, 7\rangle, \langle 6, 5, 8\rangle, \langle 7, 5, 8\rangle, \langle 7, 6, 8\rangle\}$. *Given an n-place relation, the elements of a particular ordered n-tuple of objects stand in this relation to one another if, and only if, the ordered n-tuple is a member of the extension of the relation.*

In the next section the concept of the extension of a predicate or relation will be used to analyze the conditions under which a statement whose form can be symbolized in the predicate calculus is true.

III.7 EXERCISES

A. Which of the following are the same (equal)?

1. $\{1, 2, 3\}$	4. $\langle 3, 1, 2\rangle$	7. $\{3, 2, 3\}$
2. $\{3, 2\}$	5. $\{2, \sqrt{9}, 1\}$	8. $\{1, 2, 2, 1, 3\}$
3. $\{3, 1, 2\}$	6. $\langle 1, 2, 3\rangle$	9. $\{x;\ x$ is an integer between 0 and 4$\}$

B. What are the extensions of the following predicates and relations?

1. "is a positive integer smaller than 7"
2. "is (or was) president of the United States after FDR"
3. "(1) was president after FDR, but before (2)"
4. "(1), (2), and (3) are positive integers smaller than four, and (1) is the sum of (2) and (3)"

C*. Some sets may contain other sets as members; for example, we can talk about the set of all sets having exactly two elements. However, most sets will not be members of themselves. With this in mind, consider the set of all sets that are not members of themselves. Is this set a member of itself?

III.8 TRUTH CONDITIONS

In the propositional calculus we analyzed the conditions under which formulas were true by using truth tables. Now we want to do something similar for the predicate calculus.

The first thing to observe is that we cannot talk about the truth conditions of all formulas—only of sentences. Formulas that are not sentences symbolize relations rather than statements, and only statements are true or false. For instance, we cannot ask whether the relation *(1) is the brother of (2)* is true, the question makes no sense. So it is only the truth conditions of sentences that we are analyzing. However, we cannot ask of a sentence in isolation whether it is true or false either. Sentences of the predicate calculus are only the *forms* of statements, so they are not themselves true or false. Rather, the statements whose forms are symbolized by the sentences are true or false. What must be done is to determine under what conditions a statement whose form is symbolized by a particular sentence is true. It is not enough to give a sentence by itself and ask whether it is true; rather, we must give both a sentence and an interpretation of its sentential letters, relation symbols, and individual constants, and then ask whether the statement that this sentence symbolizes when its parts have this meaning is true. Thus we cannot simply ask whether the sentence $(x)(Fx \supset Gx)$ is true, because it is only the form of a statement. But if we are given this sentence and told that F stands for "is a bachelor" and G stands for "is unmarried," then we can ask whether the statement this sentence symbolizes when interpreted in this way is true. The statement is "All bachelors are unmarried," so it is true.

It is rather awkward to talk about the truth of the statement symbolized by a sentence when interpreted in a certain way, so it is customary to talk instead about the sentence itself being true under a particular interpretation of its parts. But this should only be understood as a short way of talking about the truth of the corresponding statement.

Now let us look at the conditions under which a sentence of the predicate calculus is true under an interpretation. In the propositional calculus we found that it was unnecessary to know the *meanings* of the atomic parts of a formula in order to determine the truth value of the formula—it was only necessary to know the *truth values* of the atomic parts. Something similar is true in the predicate calculus, consider the atomic sentence *Fab*. It is not necessary to know the meaning of F, a, and b in order to determine whether this is true. It is only necessary to know the extension of F and to know what objects are referred to or to know the denotations of a and b. Suppose we know that a denotes John F. Kennedy and b denotes Harry Truman, and the extension of F is the set of ordered pairs {⟨Grover Cleveland, Eleanor Roosevelt⟩, ⟨John F. Kennedy, Harry Truman⟩, ⟨Babe Ruth, Marlene Dietrich⟩}. Then the ordered pair $\langle a, b\rangle$ is the ordered pair ⟨Kennedy, Truman⟩, and this ordered pair is in the extension of F. Therefore a and b stand in the relation F to one another, and so *Fab* is true. And we have established this without having any idea what the meaning of F might be in order to give it this peculiar extension.

Much the same is true of sentences of arbitary complexity; that is, given the universe of discourse, the extensions of all relation symbols occurring in the sentence, the truth values of all sentential letters, and the denotations of all individual constants, we can determine the truth value of the sentence itself.

Now let us try to make precise the rules we use in determining whether a sentence is true under a given interpretation. First we must make the concept of an interpretation precise. Let us say that an *interpretation* of a sentence is any assignment of the following:

1. A set of individuals to constitute the universe of discourse (this set is called the *domain* of the interpretation); and
2. denotations in the domain for the individual constants occurring in the sentence; and
3. extensions in the domain for the relation symbols occurring in the sentence; and
4. truth values for the sentential letters occurring in the sentence.[5]

The fact that a denotation *in the domain* must be assigned to each individual constant means that the object denoted must be chosen from the domain of the interpretation. And similarly, the fact that an interpretation assigns an extension *in the domain* to each relation symbol means that the objects in the extension of the relation symbol must be chosen from the domain. For example, the extension assigned to a two-place relation must be a set of ordered pairs of objects in the domain. This requirement should be emphasized, because it is a common mistake to overlook it. If it were not required that the denotations and extensions be chosen from the domain, our sentences would not have their intended meaning. For example, the sentence $[(x)Fx \supset Fa]$ means "If everything in the domain, *including a*, is *F*, then *a* is *F*." So we must ensure that the denotation of *a* is in the domain.[6]

Notice that whenever we are given a universe of discourse and the meanings of the relation symbols and individual constants in a sentence we can construct a corresponding interpretation. The domain will be

[5] Notice that an interpretation of one sentence may contain an interpretation of another sentence; for example, an interpretation of $(x)[Fx \supset Hb]$ contains an interpretation of *Hb*.

[6] Actually, in the case of relation symbols, we could assign extensions that go outside the domain if we wanted to, but we do not because any part of the extension that is outside the domain would be irrelevant to the truth value of the sentence. For example, in determining whether $(x)Fx$ is true, we must know whether every object in the domain is a member of the extension of *F*, but members of the extension of *F* that are outside the domain would be irrelevant. Thus we make the restriction that in an interpretation the extension assigned to a relation symbol must be taken from the domain.

the universe of discourse, and the extensions and denotations assigned to the relation symbols and individual constants are those determined by their meaning. In constructing an interpretation of a sentence, we simply abstract from the meaning of the sentence those features that are necessary in order to determine the truth value of the sentence.

Rules for determining the truth value of a sentence under an interpretation can now be made precise. The rules for everything but quantifiers are quite obvious:

1. An atomic sentence that is a sentential letter is true under an interpretation if, and only if, the interpretation assigns the truth value "true" to it.
2. An atomic sentence which is an *n*-place relation symbol followed by *n* individual constants is true under an interpretation if, and only if, the ordered *n*-tuple consisting of the denotations of the individual constants is a member of the extension of the relation symbol; or if $n = 1$ (so the relation symbol is a predicate letter), then the atomic sentence is true under the interpretation if, and only if, the denotation of the individual constant written after it is a member of the extension of the relation symbol.
3. $\sim P$ is true under an interpretation if, and only if, *P* is false under that interpretation.
4. $(P \& Q)$ is true under an interpretation if, and only if, both *P* and *Q* are true under that interpretation.
5. $(P \vee Q)$ is true under an interpretation if, and only if, either *P* or *Q* (or both) are true under that interpretation.
6. $(P \supset Q)$ is true under an interpretation if, and only if, either *P* is false under that interpretation, or *Q* is true under that interpretation.
7. $(P \equiv Q)$ is true under an interpretation if, and only if, *P* and *Q* have the same truth values under that interpretation.

Using Rules 1–8 we can determine the truth value under an interpretation of any sentence that does not contain any quantifiers. For example, suppose we want to know the truth value of the sentence $[P \equiv (Qa \& \sim Fab)]$ under the interpretation that assigns the domain $\{1, 2, 3\}$, the denotation 1 to *a* and the denotation 2 to *b*, the extension $\{2, 3\}$ to *Q*, and the extension $\{\langle 1, 2\rangle, \langle 2, 3\rangle\}$ to *F*, and the truth value "true" to *P*. The by Rule 1, *P* is true under the interpretation. The denotation of *a* is 1, and this is not a member of the extension of *Q*, so by Rule 2, *Qa* is false under the interpretation. The denotation of *a* is 1, and the denotation of *b* is 2, so the ordered pair consisting of the denotations of a and b is $\langle 1, 2\rangle$. This ordered pair is a member of the extension of F, so by Rule 2 Fab is true under the interpretation. Then by Rule 3, $\sim Fab$ is false under the interpretation. Both Qa and $\sim Fab$ are false, so by Rule 4, $(Qa \;\&\; \sim Fab)$ is false under the interpretation. P is true, so then by Rule 6 the whole sentence $[P \equiv (Qa \;\&\; \sim Fab)]$ is false under the interpretation.

Rules 1 and 3–7 are the same as the corresponding rules for the propositional calculus. Rule 2 just makes precise the connection between the extension of a relation and the conditions under which an ordered n-tuple of objects stand in that relation to one another.

We have yet to construct rules for determining whether universal generalizations and existential generalizations are true. These rules are more complicated than the above. To see how these rules must read, let us begin by looking at a simple case, the formula $(x)Fx$. Roughly speaking, this formula is true if, and only if, F is true of everything in the domain; that is, $(x)Fx$ says that everything in the domain is an F. One way of making this more precise is the following: $(x)Fx$ is true if, and only if, Fa would be true *regardless* of what (in the domain) we let the denotation of a be. If the domain is $\{1, 2, 3\}$, then $(x)Fx$ is true if, and only if, each of 1, 2, and 3 are in the extension of F. And the latter will be true provided that Fa would be true regardless of which of 1, 2, or 3 we assign as the denotation of a.

In order to formulate our truth rule for universal generalizations, we must have some way of varying the denotations of individual constants so that we have a way of saying that no matter what we let the denotation of a constant in some sentence be, the sentence will remain true. Given an interpretation of a sentence which does not assign a denotation to the individual constant a, let us say that another interpretation is an *a-variant* of the first interpretation if, and only if, the second interpretation is exactly like the first interpretation except that it does assign a denotation to a. Then to talk about different a-variants of an interpretation is just a way of letting the denotation of a vary over the different elements of the domain. If we consider the interpretation of $(x)Fx$ which assigns the domain $\{1, 2, 3\}$, and assigns the extension $\{1, 3\}$ to F, there will be three possible a-variants of this interpretation. These will be those interpretations that assign the same domain to the quantifiers and the same extension to F, and that assign the denotations 1, 2, and 3 to a. Then to say that Fa would be true regardless of what we take the denotation of a to be means that Fa is true under each of the a-variants of the original interpretation.

We can also talk about b-variants of an interpretation, c_{13}-variants, and so on for each individual constant. These will be those interpretations that are exactly like the initial interpretation except that they also assign a denotation to the individual constant in question.

Now we can formulate a precise truth rule for universal generalizations:

8. A universal generalization $(x)P$ is true under an interpretation if, and only if, when we choose some individual constant to which the interpretation does not assign a denotation (suppose we choose a), then $Sb\binom{a}{x}P$ is true under every a-variant of the initial interpretation.[7]

That $Sb\binom{a}{x}P$ is true under every a-variant of the initial interpretation is simply another way of saying that $Sb\binom{a}{x}P$ would be true regardless of what we let the denotation of a be.

Some examples follow of the use of this rule to determine whether sentences of the predicate calculus are true under interpretations. Consider first the sentence $(x)(Fx \supset Gx)$ and the interpretation in which the domain is the set of integers $\{1, 2, 3\}$, the extension assigned to G is $\{1, 2\}$, and the extension assigned to F is $\{2\}$. By Rule 8, $(x)(Fx \supset Gx)$ is true under this interpretation if, and only if, $(Fa \supset Ga)$ is true under every a-variant of this interpretation. There are three things in the domain, so there are three a-variants—we can let the denotation of a be either 1, or 2, or 3. So let us consider each one separately and show that $(Fa \supset Ga)$ is true under every one of them. First let the denotation of a be 1. Ga is true under that interpretation if, and only if, $1 \in \{1, 2\}$. This is the case, so Ga is true. Fa is true under this a-variant if, and only if, $1 \in \{2\}$, which is not the case. So Fa is false. Then, by Rule 6 $(Fa \supset Ga)$ is true under this a-variant. Next consider the a-variant that assigns the denotation 2 to a. Fa is true under that a-variant because $2 \in \{2\}$, and Ga is true under that a-variant because $2 \in \{1, 2\}$. So by Rule 6, $(Fa \supset Ga)$ is true under the second a-variant. The third (and last) a-variant assigns 3 to a. Both Fa and Ga are false under this a-variant, because it is not the case that $3 \in \{2\}$, and it is not the case that $3 \in \{1, 2\}$. Again by Rule 6, $(Fa \supset Ga)$ is true. Therefore, $(Fa \supset Ga)$ is true under each a-variant of our initial interpretation, and so by Rule 8, $(x)(Fx \supset Gx)$ is true under that interpretation.

[7] Notice that it makes no difference which individual constant we choose as long as it is not assigned a denotation by the original interpretation.

Let us now determine the truth value of the sentence

$$[(x)Hx \equiv (P \supset \sim(x) \sim Hx)]$$

under the interpretation that assigns the domain $\{1, 2\}$, and assigns "true" to P, and the extension $\{1\}$ to H. First evaluate the truth value of $(x)Hx$. $(x)Hx$ is true under our interpretation if, and only if, Ha is true under each a-variant of the interpretation. There are two a-variants—that assigning 1 to a, and that assigning 2 to a. Ha is true under the first a-variant because $1 \in \{1\}$, but Ha is false under the second a-variant because it is not the case that $2 \in \{1\}$. Thus Ha is not true under every a-variant of the interpretation and so $(x)Hx$ is not true under the initial interpretation. Thus the left side of the biconditional is false. Next consider $(x) \sim Hx$. This is true under the interpretation if, and only if, $\sim Ha$ is true under each a-variant of the interpretation. But by Rule 3, $\sim Ha$ is true under an a-variant if, and only if, Ha is false under that a-variant. We have seen that Ha is true under the first a-variant and false under the second, so $\sim Ha$ is false under the first a-variant and true under the second. Therefore $(x) \sim Hx$ is not true under the initial interpretation. By Rule 1, P is true, so then by Rule 6, $(P \supset \sim(x) \sim Hx)$ is true under the interpretation. Hence we have a biconditional the left side of which is false, and the right side of which is true, so the biconditional is false under our interpretation according to Rule 7.

It should be mentioned that interpretations will frequently be chosen which assign sets of integers as domains. This is only because integers are convenient to work with. These interpretations do not have any privileged status among interpretations other than their convenience. We could in principle use any set of objects for the domain of an interpretation.

Next a truth rule for existential generalizations must be constructed. The rule here is completely analogous to the case of universal generalizations. We want a sentence of the form $(\exists x)Fx$ to be true if, and only if, there is something in the domain which is an F; that is, if, and only if, there is at least one thing in the domain such that, if we choose it as the denotation of a, Fa will be true. So our rule is the following:

9. An existential generalization $(\exists x)P$ is true under an interpretation if, and only if, when we choose some individual constant to which the interpretation does not assign a denotation (suppose we choose a), then $Sb(\frac{a}{x})P$ is true under at least one a-variant of the interpretation.

Consider some examples of the use of Rule 9. Let us calculate the truth value of $(\exists x)Bxa$ under the interpretation that assigns the set of all past presidents of the United States as the domain, assigns George Washington as the denotation of a, and assigns to B the set of all ordered pairs of past presidents such that the first member of each ordered pair was president before the second member; thus B is the relation "(1) was president before (2)." To determine the truth value of $(\exists x)Bxa$ under this interpretation we must first choose an individual constant to which the interpretation does not assign a denotation. We cannot choose a, because it occurs in Bxa and so has been assigned a denotation, so we choose b. Then $(\exists x)Bxa$ is true under this interpretation if, and only if, Bba is true under at least one b-variant of this interpretation. But there is no denotation we can assign to b making Bba true, because such a denotation would have to be a president before Washington, and there was none. Therefore Bba is not true under any b-variant of our interpretation, and so $(\exists x)Bxa$ is false under the original interpretation.

Now here is a more complicated example in which we have one quantifier occurring within the scope of another quantifier. Consider the sentence $(x)(\exists y)Lxy$ and the interpretation that assigns the domain $\{1, 2, 3\}$ and in which L is the "less than" relation; that is, the extension assigned to L is $\{\langle 1, 2\rangle, \langle 1, 3\rangle, \langle 2, 3\rangle\}$. $(x)(\exists y)Lxy$ is true under this interpretation if, and only if, $(\exists y)Lay$ is true under every a-variant of this interpretation. There are three a-variants—those in which we assign either 1, or 2, or 3 to a. But now, $(\exists y)Lay$ is true under a given a-variant if, and only if, Lab is true under some b-variant of that a-variant. For each a-variant there are three b-variants of it—those assigning 1, 2, or 3 to b. We can diagram the whole thing as follows:

	$b = 1$	Lab false		
$a = 1$	$b = 2$	Lab true	$(\exists y)Lay$ true	
	$b = 3$	Lab true		
	$b = 1$	Lab false		
$a = 2$	$b = 2$	Lab false	$(\exists y)Lay$ true	$(x)(\exists y)Lxy$ false
	$b = 3$	Lab true		
	$b = 1$	Lab false		
$a = 3$	$b = 2$	Lab false	$(\exists y)Lay$ false	
	$b = 3$	Lab false		

Lab is true under the second and third *b*-variant of the first *a*-variant, so $(\exists y)Lay$ is true under the first *a*-variant. *Lab* is true under the third *b*-variant of the second *a*-variant, so $(\exists y)Lay$ is also true under the second *a*-variant. But *Lab* is not true under any *b*-variant of the third *a*-variant, so $(\exists y)Lay$ is false under the third *a*-variant. Therefore, it is not true that $(\exists y)Lay$ is true under every *a*-variant of the initial interpretation, and so $(x)(\exists y)Lxy$ is false under the initial interpretation.

It should be observed that it will often be unnecessary to work through all of the steps of the truth rules in order to determine whether a sentence is true under a given interpretation. If we can see what a sentence means under an interpretation we can often see immediately whether it is true. Recall the above interpretation of $(\exists x)Bxa$. Under that interpretation, $(\exists x)Bxa$ means "There was someone who was president before Washington." Seeing this, we know immediately that the sentence is false. Or consider the interpretation of the sentence $(x)(\exists y)Lxy$ in the preceding example. Under that interpretation this sentence means "For every integer there is another integer such that the first is smaller than the second," that is, "For every integer there is a larger integer." If the domain had been the set of *all* integers this would have been true, but as the domain is just $\{1, 2, 3\}$ it is false—there is no integer in the domain larger than 3. Clearly, the use of truth rules is often rather cumbersome and unnecessary. However, we must have such rules for those cases in which the meaning of a sentence is so complicated that we cannot tell simply by reflection whether it is true.

III.8 EXERCISES

Let the universe of discourse be the set $\{1, 2, 3\}$, and let *G* mean "is greater than one," *F* mean "is less than three," *H* mean "(1) is not equal to (2)," *I* mean "(1) is equal to (2)," *L* mean "(1) is less than (2)," *P* mean "one is not equal to two," and let *a* denote 1, *b* denote 2, and *c* denote 3. This generates the interpretation that assigns the domain $\{1, 2, 3\}$ to the quantifiers, the extensions $\{1, 2\}$ to *F*, $\{2, 3\}$ to *G*, $\{\langle 1, 2\rangle, \langle 1, 3\rangle, \langle 2, 1\rangle, \langle 2, 3\rangle, \langle 3, 1\rangle, \langle 3, 2\rangle\}$ to *H*, $\{\langle 1, 1\rangle, \langle 2, 2\rangle, \langle 3, 3\rangle\}$ to *I*, and $\{\langle 1, 2\rangle, \langle 1, 3\rangle, \langle 2, 3\rangle\}$ to *L*, and assigns "true" to *P*, 1 to *a*, 2 to *b*, and 3 to *c*. Under this interpretation, express the meaning of each of the following sentences in idiomatic English and determine whether it is true:

1. Fa
2. $[(Fa \& Fb) \& Fc]$
3. $(x)Fx$
4. $[(x)Gx \equiv ((Ga \& Gb) \& Gc)]$
5. $(\exists x)Gx$
6. $(Iab \vee \sim Iab)$
7. $(Iab \vee \sim Hab)$
8. $(x)(\exists y)Ixy$
9. $((x)Ixx \& (x) \sim Hxx)$
10. $(\exists y)(x)Ixy$
11. $(x)(y)(Ixy \supset \sim Hxy)$
12. $(\exists x)(y)(Lxy \vee Ixy)$
13. $(x)(y)(Lxy \supset \sim Ixy)$
14. $(\exists x)(y)(Fx \supset Fy)$
15. $(\exists y)(x)(Lxy \vee Icx)$
16. $(P \equiv (\exists x) \sim Ixx)$
17. $(x)(y)(z)[(Lxy \& Lyz) \supset Lxz]$
18. $(x) \sim Lxx$
19. $(x)(y)(Lxy \supset \sim Lyx)$

III.9 VALIDITY AND FORMAL ANALYTICITY

In the propositional calculus we defined the notion of tautologicity, and showed that a formula of the propositional calculus was a tautology if, and only if, it was the form of a formally analytic statement. We would now like to find out how to define an analogous notion for formulas of the predicate calculus.

The most natural suggestion is that a sentence of the predicate calculus is formally analytic if, and only if, it is true under every interpretation. But this suggestion runs into a difficulty. Suppose we have the sentence $(x)(Fx \supset Gx)$, which does not contain any individual constants. Then, if we look at the definition of an interpretation, we see that there is nothing to prevent us from taking the domain of the interpretation to be the empty set. On the other hand, if we have a sentence like $(Fa \supset Ga)$, which does contain occurrences of individual constants, we cannot take the domain of an interpretation to be the empty set, because we must be able to assign a member of the domain as the denotation of the individual constants (and the empty set has no members). This immediately leads to a difficulty for our definition of truth under an interpretation. Our rule for determining the truth value of universal generalizations and existential generalizations is not applicable in an interpretation having an empty domain, because such an interpretation has no *a*-variants; that is, we cannot assign a denotation to *a* in such an interpretation because the domain has no members. Thus our definition of truth only works for interpretations having nonempty domains, and so it is not even clear what it *means* to say that a universal or existential generalization is true under every interpretation.

Therefore, we cannot use "true under every interpretation" as an analogue of formal analyticity, because we have not explained what it means to say that a sentence is true under an interpretation with an empty domain. We might try to rectify this difficulty by adopting some convention concerning the empty domain, such as saying that a universal generalization is always true and an existential generalization is always false under an interpretation having an empty domain. However, this is not the customary way of proceeding. Instead, let us examine another possibility.

Because of the difficulties concerning the empty domain, we might instead suggest that the analogue of formal analyticity should be "true under every interpretation having a nonempty domain." This is in fact what is generally done. We say that a sentence of the predicate calculus is *valid* if, and only if, it is true under every interpretation having a nonempty domain. If a formula, say P, is valid, we will write $\models P$. But given this definition, there is some difficulty in relating validity to formal analyticity. The difficulty concerns our practice of restricting the universe of discourse (the domain of the quantifiers) in symbolizing statements. Given that we can restrict the universe of discourse to any set we like, it seems that we can have valid sentences that are the forms of false statements.

Let us suppose that we take as the domain for our quantifiers the set of unicorns, which is the empty set. Letting R stand for "is red," we will symbolize the statement "There is a unicorn which is either red or not red" as $(\exists x)(Rx \vee {\sim}Rx)$. The statement "There is a unicorn which is either red or not red" is false, because there are no unicorns. But the sentence $(\exists x)(Rx \vee {\sim}Rx)$ is valid; that is, true under every interpretation having a nonempty domain. This is because $(\exists x)(Rx \vee {\sim}Rx)$ is true under an interpretation having a nonempty domain if, and only if, $(Ra \vee {\sim}Ra)$ is true under some a-variant of that interpretation. But in fact, $(Ra \vee {\sim}Ra)$ will be true no matter what we assign as the denotation of a. So $(Ra \vee {\sim}Ra)$ will be true under every a-variant of the interpretation, and $(\exists x)(Rx \vee {\sim}Rx)$ will be true under the interpretation. Therefore, $(\exists x)(Rx \vee {\sim}Rx)$ is valid, but it is the form of a false statement. But if a statement is false it is certainly not formally analytic, and so a valid sentence need not be the form of a formally analytic statement.

The above difficulty arises from restricting the universe of discourse to a limited domain. In order to avoid this difficulty, we might require that in symbolizing the form of a statement we always use the universal set (the set of everything) for the universe of discourse. Let us see if this avoids our difficulty. If a sentence of the predicate calculus is true under every interpretation, then in particular it will be true under every interpretation whose domain is the universal set. Therefore, if this sentence is what we obtain when we symbolize a statement (without using a restricted universe of discourse), then that statement must be true. The statement is *made* true by its logical form, so it is formally analytic. We can conclude then that *if a sentence is valid, then any statement having that form is formally analytic.*

Now it must be seen whether the converse is true. Is it true that, if a statement can be symbolized in the predicate calculus, and it is formally analytic, then it is valid? At first there seems to be no reason for thinking that this is true. We *do* have the following: if a sentence results from symbolizing the form of a formally analytic statement, then that sentence must be true under every interpretation whose domain is the universal set. This is because every such interpretation must correspond to another statement having the same logical form, and since the original statement is formally analytic, every statement having that logical form must be true. But this only shows that the sentence must be true under every interpretation having this one fixed domain. In order to show that the sentence is valid we must show that it is true under *every* interpretation having *any* nonempty domain. Fortunately, this can be shown. There is a famous theorem in logic called the Löwenheim Theorem, which says:

> If a sentence of the predicate calculus is true under every interpretation having some particular infinite domain, then the sentence is true under every interpretation having any nonempty domain (that is, the sentence is valid).[8]

The universal set is clearly infinite, because among other things it contains all numbers. Therefore, by the Löwenheim Theorem, if a sentence is true under every interpretation whose domain is the universal set, then the sentence is valid. And therefore, *if a sentence results from symbolizing the form of a formally analytic statement, then the sentence is valid.*

Combining the above two results, we have the following: *If the form of a statement can be expressed as a sentence of the predicate calculus, then the statement is formally analytic if, and only if, the sentence is valid.* Thus validity provides our desired analogue for formal analyticity.

[8] Why this theorem is true will be seen in Section III.16.

Unfortunately, validity provides an analogue for formal analyticity only at the expense of our being unable to use restricted universes in symbolizing the forms of statements. This was a very useful shortcut. There is, however, a way that it can be salvaged. Just as in the propositional calculus, the concept of validity will be used for assessing the soundness of argument forms by seeing whether their corresponding conditionals are valid. Now, suppose that we have an argument form that we have symbolized using a restricted universe of discourse; for example:

Everyone who has a father has a grandfather.
Everyone who has a grandfather has a grandmother.

Therefore, everyone who has a father has a grandmother.

symbolized as

$$\begin{array}{l} (x)(Fx \supset Gx) \\ (x)(Gx \supset Mx) \\ \hline (x)(Fx \supset Mx) \end{array}$$

by using the class of people as a restricted universe of discourse. Suppose then that we formed the corresponding conditional

$$\{[(x)(Fx \supset Gx) \ \& \ (x)(Gx \supset Mx)] \supset (x)(Fx \supset Mx)\}$$

and ascertained somehow that it is valid. It does not follow from this that the initial argument is sound. This is because should the universe of discourse be empty—should there be no people in the world—the corresponding conditional of the initial argument might even be false (much less formally analytic), while its symbolization is still valid. However, it does follow that another argument, which we obtain from the initial argument by adding an additional premise, is sound:

There exists at least one person.
Everyone who has a father has a grandfather.
Everyone who has a grandfather has a grandmother.

Therefore, everyone who has a father has a grandmother.

Here we have just added the premise that the universe of discourse is nonempty. Why is this new argument sound? Suppose that the conditional $\{[(x)(Fx \supset Gx) \ \& \ (x)(Gx \supset Mx)] \supset (x)(Fx \supset Mx)\}$ is valid.

Then it is true under every interpretation having a nonempty domain. Therefore, *if* the class of people is nonempty, then

$$\{[(x)(Fx \supset Gx) \ \& \ (x)(Gx \supset Mx)] \supset (x)(Fx \supset Mx)\}$$

is true under every interpretation whose domain is the class of people. Therefore, *if* the class of people is nonempty, then if the two premises of the original argument are true so must the conclusion be (because these statements correspond to one interpretation of the conditional). But the new premise we have added to the original argument *is* the premise that the class of people is nonempty. Therefore, given *all three* of these premises, it follows that the conclusion must be true, and so the new three-premise argument is sound.

Generally, given an argument which we symbolize using a restricted domain, it does not follow from the validity of the corresponding conditional that the argument is sound. But it *does* follow from the validity of the corresponding conditional that a new argument, which we construct by adding the additional premise that there is at least one thing in the domain, is sound. Generally there will be no question about this new premise being false, and so we can confidently use the argument to conclude that if the original premises are true then the conclusion is true. However, if there is some doubt about whether the domain is nonempty, then in order to infer the consequent of the original argument from its premises, we must symbolize the argument all over again without using a restricted universe of discourse.

Because of the difficulties with the empty domain, the customary procedure is to redefine "interpretation" to require that an interpretation always have a nonempty domain. Henceforth, when we write "interpretation" we mean "interpretation having a nonempty domain."

Related to the concept of validity are three other concepts. We say that a sentence is *invalid* if, and only if, it is not valid. And we say that a sentence is *consistent* (often called *satisfiable*) if, and only if, its negation is invalid. A sentence is *inconsistent* if, and only if, it is not consistent. Thus a sentence is invalid if, and only if, it is false under at least one interpretation; a sentence is consistent if, and only if, it is true under at least one interpretation; and a sentence is inconsistent if, and only if, it is false under every interpretation. Notice that a sentence is inconsistent if, and only if, it implies an explicit contradiction of the form $(Q \ \& \ {\sim}Q)$. This is because a formula P implies $(Q \ \& \ {\sim}Q)$ if, and only if, $[P \supset (Q \ \& \ {\sim}Q)]$ is valid. By E4 this is equivalent to

$$[{\sim}P \vee (Q \ \& \ {\sim}Q)]$$

and by E17 this is equivalent to $\sim P$. Thus $\sim P$ is valid (P is inconsistent) if, and only if, P implies an explicit contradiction.

In order to show that a sentence is consistent it suffices to find one interpretation under which it is true, and in order to show that a sentence is invalid it suffices to find one interpretation under which it is false. But to show that a sentence is valid you must show that it is true under *every* interpretation, and in order to show that a sentence is inconsistent you must show that it is false under *every* interpretation. Thus, to establish either that a sentence is valid or that it is inconsistent, you must give some sort of general argument about all interpretations. However, to establish that a sentence is consistent, or that it is invalid you can merely give an example of an interpretation under which it is true, or false, respectively. This must be emphasized: *you cannot establish that a sentence is valid (or that it is inconsistent) by looking at a single interpretation, although you can establish that it is invalid (or that it is consistent) by looking at a single interpretation.*

Let us look at some examples. Suppose we want to show that $(x)(Fx \supset Gx)$ is invalid. Then we must find an interpretation under which it is false. Since it is much easier to work with interpretations having small domains, let us try to find an interpretation under which $(x)(Fx \supset Gx)$ is false, which has a one-element domain. Suppose the domain is $\{1\}$. In order to make $(x)(Fx \supset Gx)$ false under some interpretation having this domain, we must make $(Fa \supset Ga)$ false under some a-variant of that interpretation. But there will only be one a-variant, because there is only one object in the domain. So the denotation assigned to a must be 1. Then in order for $(Fa \supset Ga)$ to be false, we must make Fa true and Ga false. Thus, 1 must be a member of the extension of F, but not of G. This means that we must choose the extension $\{1\}$ for F and the extension $\varnothing$ for G (because if 1 is not to be in the extension of G, then nothing can be because there is nothing else in the domain). Thus, $(x)(Fx \supset Gx)$ is false under the interpretation whose domain is $\{1\}$ and that assigns the extension $\{1\}$ to F and $\varnothing$ to G. Therefore, $(x)(Fx \supset Gx)$ is invalid.

Next let us show that $(x)(Fx \supset Gx)$ is consistent. To do this we must find an interpretation under which it is true. Again let us see if we can do it with the simplest possible domain—a one-element domain, using the domain $\{1\}$ again. In order for $(x)(Fx \supset Gx)$ to be true under an interpretation having this domain, $(Fa \supset Ga)$ must be true under every a-variant of the interpretation. But once again there will only be one a-variant—that assigning 1 to a. So, when the denotation of a is

1, we must have $(Fa \supset Ga)$ true. There are a number of ways we could do this. We can either have both *Fa* and *Ga* true, or *Fa* false and *Ga* true, or both *Fa* and *Ga* false. Let us choose the first alternative. In order to make both *Fa* and *Ga* true, the extensions of both must contain 1, and their extensions cannot contain anything else, because there is nothing else in the domain. So the extensions must be simply {1}. Therefore, $(x)(Fx \supset Gx)$ is true under the interpretation whose domain is {1}, and that assigns the extension {1} to both *F* and *G*.

To show that $(x)(\exists y)(Hxy \;\&\; \sim Hyy)$ is consistent, we must find an interpretation under which it is true. Again we can start off by seeking an interpretation having the one-element domain {1}. Then in order for $(x)(\exists y)(Hxy \;\&\; \sim Hyy)$ to be true under an interpretation having this domain, $(\exists y)(Hay \;\&\; \sim Hyy)$ must be true under every *a*-variant of the interpretation. But there is only one *a*-variant, that assigning 1 to *a*. So $(\exists y)(Hay \;\&\; \sim Hyy)$ must be true under the *a*-variant of the interpretation that assigns 1 to *a*. But $(\exists y)(Hay \;\&\; \sim Hyy)$ is true under that *a*-variant if, and only if, $(Hab \;\&\; \sim Hbb)$ is true under at least one *b*-variant of that *a*-variant. But again, there is only one *b*-variant, that assigning 1 to *b*. So $(Hab \;\&\; \sim Hbb)$ must be true under some interpretation that assigns 1 to both *a* and *b*. But this is impossible, because if *a* and *b* have the same denotation, then *Hab* will be true if, and only if, *Hbb* is true (because it is the same ordered pair in both cases), and so $(Hab \;\&\; \sim Hbb)$ cannot be true. This shows that there cannot be an interpretation having a one-element domain in which

$$(x)(\exists y)(Hxy \;\&\; \sim Hyy)$$

is true.

Let us see if we can find an interpretation having a two-element domain, {1, 2}, in which $(x)(\exists y)(Hxy \;\&\; \sim Hyy)$ is true. This sentence will be true under an interpretation having this domain if, and only if, $(Hab \;\&\; \sim Hbb)$ is true under some *b*-variant of each *a*-variant of the interpretation. There are two *a*-variants—those assigning 1 and 2 to *a*—and there are two *b*-variants of each *a*-variant—those assigning 1 and 2 to *b*. Is there an extension we can assign to *H* that will give this result? First consider the *a*-variant assigning the denotation 1 to *a*. Under some *b*-variant of this *a*-variant $(Hab \;\&\; \sim Hbb)$ must be true. For $(Hab \;\&\; \sim Hbb)$ to be true we must have *Hab* true and *Hbb* false. To have *Hab* true under some *b*-variant of *a*-variant, there must be some ordered pair in the extension of *H* whose first element is 1. The second element must either be 1 or 2, depending upon which denotation b has. If we allow Hab to be true under the b-variant that assigns 1 to b, then $\langle 1, 1\rangle$ will be a member of the extension of H. But then Hbb will also be true, and $(Hab \ \& \ {\sim}Hbb)$ will then be false; so this b-variant will not do the trick. Suppose then that we let Hab be true under the other b-variant—that assigning 2 to b. Then $\langle 1, 2\rangle$ must be in the extension of H. Also, in order to have Hbb false, $\langle 2, 2\rangle$ must not be in the extension of H. Thus, by requiring that $\langle 1, 2\rangle$ is in the extension of H, and $\langle 2, 2\rangle$ is not in the extension of H, we can make $(Hab \ \& \ {\sim}Hbb)$ true in some b-variant of the first a-variant. Now let us see if we can also make it true in the second a-variant. Here we assign 2 to a. Again, we want to make $(Hab \ \& \ {\sim}Hbb)$ true under some b-variant. Let us see whether we can make it true under the b-variant that assigns 1 to b. In order to make Hab true and Hbb false under that b-variant of this a-variant, we must have $\langle 2, 1\rangle$ in the extension of H, and we must not have $\langle 1, 1\rangle$ in the extension of H. Therefore, we see that we can make $(x)(\exists y)(Hxy \ \& \ {\sim}Hyy)$ true under an interpretation whose domain is $\{1, 2\}$ if we can assign an extension to H, which contains $\langle 1, 2\rangle$ and $\langle 2, 1\rangle$, but does not contain $\langle 1, 1\rangle$ or $\langle 2, 2\rangle$. We can do this simply by assigning the extension $\{\langle 1, 2\rangle, \langle 2, 1\rangle\}$; therefore, $(x)(\exists y)(Hxy \ \& \ {\sim}Hyy)$ is consistent.

It is generally much easier to work with interpretations having small domains, so when trying to find an interpretation under which a sentence is false, it is generally wise to begin with the smallest possible domain, a one-element domain. If this does not work, then try a two-element domain, and so on. Occasionally, however, a sentence will only be true under an interpretation having an infinite domain. The sentence

$$\{(x)(y)(z)[(Rxy \ \& \ Ryz) \supset Rxz] \ \& \ [(x) \sim Rxx \ \& \ (x)(\exists y)Rxy]\}$$

can be shown to be false under any interpretation having a finite domain. But it is consistent, because it can be made true in an interpretation having an infinite domain. If we let the domain be the set of all positive integers, and let R be the relation "(1) is a positive integer less than (2)," this sentence is true. So if one cannot find an interpretation having a small domain under which a sentence is true, it does not automatically follow that the sentence is inconsistent. You might have to try an infinite domain in order to make the sentence true.

This last example also illustrates something else. Thus far, in trying to find interpretations under which a sentence is true, we have more or less stumbled around until we found something that works. But there is another technique that often works better. As in the above example, we might try to think of meanings we can assign to the relation symbols and individual constants in a sentence, which would make it symbolize a true statement. For example, $(x)(Fx \supset Gx)$ means "Every F is a G." If we want to make this true we might let F stand for "is a bachelor" and G stand for "is unmarried." Then $(x)(Fx \supset Gx)$ will symbolize the true statement "All bachelors are unmarried." Thus, $(x)(Fx \supset Gx)$ is true under the interpretation whose domain is the set of people, and which assigns the set of bachelors as the extension of F and the set of unmarried people as the extension of G. On the other hand, if we want to make $(x)(Fx \supset Gx)$ false, we might let F stand for "is a woman" and G for "has a beard." $(x)(Fx \supset Gx)$ is false under that interpretation whose domain is the set of people, and which assigns the set of women as the extension of F and the set of bearded people as the extension of G.

A more complicated example is $(x)(\exists y)(Hxy \,\&\, {\sim}Hyy)$. If we let the universe of discourse be the set of positive integers, and we let H mean "(1) is less than (2)," then $(x)(\exists y)(Hxy \,\&\, {\sim}Hyy)$ means "For every positive integer, there is a positive integer greater than it but not greater than itself." Thus $(x)(\exists y)(Hxy \,\&\, {\sim}Hyy)$ is true under that interpretation whose domain is the set of positive integers, and which assigns to H the extension $\{\langle x, y\rangle;\ x \text{ is less than } y\}$. Another meaning we might assign to H to make $(x)(\exists y)(Hxy \,\&\, {\sim}Hyy)$ true is "(1) is not equal to (2)." Then $(x)(\exists y)(Hxy \,\&\, {\sim}Hyy)$ means "Given any positive integer, there is a positive integer that is unequal to it but not unequal to itself." Notice that if we let H mean this, and we let our universe of discourse be $\{1, 2\}$, we arrive at the same interpretation as we did on page 130 merely by considering the truth rules.

We have seen that we can show a sentence to be consistent, or invalid, simply by finding a single interpretation under which it is true, or false, respectively. But it is more difficult to show that a sentence is inconsistent or valid. In doing this we cannot work with a single interpretation. We must give some sort of general argument to show that the sentence is either true under all interpretations (valid), or else false under all interpretations (inconsistent), depending upon what we are trying to show. Let us look at some examples of this.

Suppose first that we want to establish the validity of $(x)(Px \supset Px)$. This sentence is valid if, and only if, it is true under every interpretation. So consider an arbitrary interpretation. $(x)(Px \supset Px)$ is true under that interpretation if, and only if, $(Pa \supset Pa)$ is true under every a-variant of that interpretation. But $(Pa \supset Pa)$ will be true under every a-variant of any interpretation, because it could only be false if the antecedent, Pa, were true, while the consequent, Pa again, were false, which is impossible. So $(Pa \supset Pa)$ is true under every a-variant of any arbitrary interpretation. Therefore, $(x)(Px \supset Px)$ is true under the interpretation. Since this is true for *any* interpretation, $(x)(Px \supset Px)$ must be true under every interpretation; that is, $(x)(Px \supset Px)$ must be valid.

Next suppose we want to show that $(\exists y)(Fy \ \& \ {\sim}Fy)$ is inconsistent; that is, false under every interpretation. Again, consider an arbitrary interpretation. $(\exists y)(Fy \ \& \ {\sim}Fy)$ will be true under that interpretation if, and only if, $(Fa \ \& \ {\sim}Fa)$ is true under some a-variant of that interpretation. But in order for $(Fa \ \& \ {\sim}Fa)$ to be true, Fa would have to be both true and false, which is impossible. Thus, $(Fa \ \& \ {\sim}Fa)$ is not true under any a-variant of the interpretation, and $(\exists y)(Fy \ \& \ {\sim}Fy)$ is false under the interpretation. This is true for any interpretation we might choose, so $(\exists y)(Fy \ \& \ {\sim}Fy)$ will be false under every interpretation, and so inconsistent.

To take a more complicated case, let us show that $(\exists x)(Fx \supset (y)Fy)$ is valid. Consider any interpretation. $(\exists x)(Fx \supset (y)Fy)$ will be true under that interpretation if, and only if, $(Fa \supset (y)Fy)$ is true under some a-variant of that interpretation. $(Fa \supset (y)Fy)$ will be true under a given a-variant as long as we do not have Fa true and $(y)Fy$ false. $(y)Fy$ does not contain a, so it is true under the a-variant if, and only if, it is true under the original interpretation. Now we have two cases to consider:

CASE 1. Suppose $(y)Fy$ is true under the original interpretation. Then $(Fa \supset (y)Fy)$ will be true under every a-variant of the original interpretation (because it has a true consequent), and so

$$(\exists x)(Fx \supset (y)Fy)$$

will be true under the original interpretation.

CASE 2. Suppose $(y)Fy$ is false under the original interpretation, then Fb must be false under some b-variant of the original interpretation; that is, there must be something we can assign as the denotation of b that will make Fb false. But now, suppose we consider the a-variant that assigns to a as its denotation the same thing as the denotation of b. Then Fa will also be false. But then $(Fa \supset (y)Fy)$, being a conditional with a false antecedent, will be true under that a-variant. Thus $(\exists x)(Fx \supset (y)Fy)$ will be true under the original interpretation. In either case, $(\exists x)(Fx \supset (y)Fy)$ will be true under the original interpretation. This is true for every interpretation, so $(\exists x)(Fx \supset (y)Fy)$ must be true under every interpretation—it is valid.

It is often quite difficult to give an argument to show that a sentence is valid. Clearly it would be advantageous to have some sort of general rules to help us along, like the rules of tautological implication and tautological replacement in the propositional calculus. In the next few sections we will be concerned with finding such rules, and then we will use these rules to construct derivations in the predicate calculus, much like our derivations in the propositional calculus.

III.9 EXERCISES

A. For each of the following sentences, find an interpretation under which it is true:

1. Fa	6. $(x)(\exists y)Hxy$
2. $(Fa \& \sim Fb)$	7. $(\exists y)(x)Hxy$
3. $(x)Fx$	8. $(x)(Fx \supset Gx)$
4. $(\exists x)Fxa$	9. $(\exists x)(\exists y)(Hxy \& \sim Hyx)$
5. $((\exists x)Fx \& (\exists x) \sim Fx)$	10. $(x)(Fxx \supset (\exists y)Fxy)$

B*. 1. Recalling that every formula of the propositional calculus is a sentence of the predicate calculus, show that if a formula of the propositional calculus is truth-functionally consistent, then it is consistent.

2. If we were to extend the definition of truth to interpretations having empty domains, we would rule that all universal generalizations are false. Give an example of a sentence that would then be true under every interpretation having an empty domain but false under all other interpretations.

III.10 IMPLICATION AND EQUIVALENCE

In the propositional calculus, corresponding to the concept of a tautology, we had the concepts of tautological implication and tautological equivalence. In the predicate calculus, corresponding to the concept of validity, we have the concepts of implication and equivalence. Given two sentences of the predicate calculus, we say that the first *implies* the second if, and only if, the conditional having the first sentence as its antecedent and the second sentence as its consequent is valid. So, for example, $(x)Fx$ implies Fa, because the conditional $[(x)Fx \supset Fa]$ is valid. Generally, we say that a finite *set* of sentences *implies* another sentence if, and only if, the conditional whose antecedent is the conjunction of the sentences in the set and whose consequent is the other sentence, is valid. Because all tautologies are valid, it follows that if a sentence or set of sentences tautologically implies another sentence, then it implies the other sentence; that is, tautological implication is a special case of implication in general.

If a set of sentences, say $P_1, \ldots, P_n$, implies a sentence Q, we will write $P_1, \ldots, P_n \models Q$.

To say that a conditional is true under an interpretation means that if the interpretation makes the antecedent true it also makes the consequent true. Thus to say that a conditional is valid means that every interpretation of it that makes the antecedent true also makes the consequent true. Thus, $P_1, \ldots, P_n \models Q$ if, and only if, every interpretation (that contains interpretations of all these formulas) making $P_1, \ldots, P_n$ all true also makes Q true.

We define equivalence in an analogous manner. We say that two sentences of the predicate calculus are *equivalent* if, and only if, the biconditional having these two sentences on opposite sides is valid. $(x)(Fx \supset Gx)$ is equivalent to $(x)(\sim Gx \supset \sim Fx)$, because

$$[(x)(Fx \supset Gx) \equiv (x)(\sim Gx \supset \sim Fx)]$$

is valid.

Notice that two sentences are equivalent if, and only if, each implies the other. Therefore, there is another way of defining equivalence: two sentences are equivalent if, and only if, every interpretation (containing interpretations of both formulas) that makes one true also makes the other true.

We can construct a rule of replacement for the predicate calculus that is analogous to the rule of tautological replacement for the propositional calculus: *if one sentence can be obtained from a second by replacing one of the parts of the first sentence by a sentence to which it is equivalent, then the second sentence is equivalent to the first sentence.* That this is so should be clear by thinking about the truth rules for sentences of the predicate calculus. If one sentence occurs as part of a larger sentence, then it is only the truth value of the smaller sentence that is relevant in determining the truth value of the larger sentence. Hence, if we replace the smaller sentence by another sentence to which it is equivalent the truth value of the larger sentence will remain unchanged.

Unfortunately, the above rule of replacement is not as useful as was the corresponding rule in the propositional calculus, because in the predicate calculus the parts of sentences are not generally sentences. This rule only allows us to replace parts that are sentences because equivalence is only defined for sentences. It will be seen in Section III.12 how this rule can be generalized to cover replacement of formulas that are not sentences.

At this point it is useful to list a few implications and equivalences. To begin with, all of the tautological implications I1–I15 are implications, and all of the tautological equivalences E1–E21 are equivalences. In addition we have the following implications

I16. $(x)P \;\square\!\!= \; Sb\binom{a}{x}P$

I17. $Sb\binom{a}{x}P \;\square\!\!= \; (\exists x)P$

Roughly, I16 says that if P is true of everything, then it is true of a. I17 says that if P is true of a, then it is true of something.

In addition to the tautological equivalences, we have the following equivalences:

E22.
$(x)P$ eq. $\sim(\exists x)\sim P$
$(\exists x)P$ eq. $\sim(x)\sim P$
$\sim(x)P$ eq. $(\exists x)\sim P$
$\sim(\exists x)P$ eq. $(x)\sim P$

E23. $(x)(P \vee Q)$ eq. $[(x)P \vee Q]$
E24. $(Q \supset (x)P)$ eq. $(x)(Q \supset P)$
E25. $((\exists x)P \supset Q)$ eq. $(x)(P \supset Q)$
provided there are no free occurrences of x in Q.

E26. $(x)P$ eq. $(y)Sb\binom{y}{x}P$
E27. $(\exists x)P$ eq. $(\exists y)Sb\binom{y}{x}P$
provided y does not occur in P.

These implications and equivalences should be thought of as implication and equivalence *schemes* rather than particular implications and equivalences. Any pairs of sentences that can be obtained from these by substituting other variables for x and y, other individual constants for a, and other formulas for P and Q, will also be implications or equivalences. From E24 for example, we can conclude that

$$[(x)Hx \supset (y)(Hy \vee Fy)]$$

is equivalent to $(y)[(x)Hx \supset (Hy \vee Fy)]$.

E24 and E25 are somewhat surprising, so let us see why they are true. First consider E24. By E4, $(Q \supset (x)P)$ is equivalent to $({\sim} Q \vee (x)P)$. By E12, the latter is equivalent to $((x)P \vee {\sim} Q)$, and by E23 this is equivalent to $(x)(P \vee {\sim} Q)$. $(x)(P \vee {\sim} Q)$ is true under a given interpretation if, and only if, $Sb(^{a}_{x})(P \vee {\sim} Q)$ is true under every a-variant of that interpretation (where a is some individual constant not assigned a denotation by the interpretation). By E12, $Sb(^{a}_{x})(P \vee {\sim} Q)$ is equivalent to $Sb(^{a}_{x})({\sim} Q \vee P)$, and then by E4 this is equivalent to $Sb(^{a}_{x})(Q \supset P)$. Thus $(x)(P \vee {\sim} Q)$ is true under an interpretation if, and only if, $Sb(^{a}_{x})(Q \supset P)$ is true under every a-variant of that interpretation. But the latter is just the condition under which $(x)(Q \supset P)$ is true. Therefore, $(Q \supset (x)P)$ is equivalent to $(x)(Q \supset P)$.

Similarly, $((\exists x)P \supset Q)$ is equivalent to $({\sim} (\exists x)P \vee Q)$, by E4. By E22 this is equivalent to $((x) {\sim} P \vee Q)$, which by E23 is equivalent to $(x)({\sim} P \vee Q)$. By reasoning as above and using E4, this can be seen to be equivalent to $(x)(P \supset Q)$.

It should be emphasized that in using E23–E25, Q cannot contain any free occurrences of x. If Q did contain free occurrences of x, then, for example, $((x)P \vee Q)$ would contain free occurrences of x and so would not even be a sentence.

E26 and E27 result from the fact that, for example, $(x)Fx$ and $(y)Fy$ mean the same thing. They both say simply that everything is an F. In terms of the truth rules, they are both true under an interpretation if, and only if, Fa is true under every a-variant of the interpretation. Thus, it makes no difference which bound variable we use. However, the restriction that y does not occur in P is essential. Without it we would find ourselves saying, for example, that $(x)(y)Hxy$ is equivalent to $(y)(y)Hyy$. These sentences, however, are not equivalent. The first says that given any two things, they stand in the relation H to one another, whereas the second simply says that everything stands in the relation H to itself (the first quantifier is a vacuous quantifier and does not affect the meaning).

III.10 EXERCISES

1. Show that if a sentence is implied by a valid sentence, then it is valid.
2. Show that implication is strongly transitive.
3. Show that implication is adjunctive.

4*. Using the result of 2, show that if P, Q, and R are sentences, and $P \mathrel{[\,]\!\!=} Q$, $Q \mathrel{[\,]\!\!=} R$, and $R \mathrel{[\,]\!\!=} P$, then P, Q, and R are all equivalent.

III.11 UNIVERSAL GENERALIZATION

There is another useful way to obtain valid sentences from valid sentences. This turns around the following fact: *if P is any formula of the predicate calculus which contains free occurrences of some variable, say "x," but not of any other variable, then given any individual constant, say "a," which does not occur in P, if* $Sb(\frac{a}{x})P$ *is valid, then* $(x)P$ *is valid.* For example, consider the formula $(Fx \vee \sim Fx)$. The sentence $(Fa \vee \sim Fa)$ is valid, because it is a tautology. Therefore, by the above principle, $(x)(Fx \vee \sim Fx)$ is valid. Now, why is this principle true? Suppose $Sb(\frac{a}{x})P$ is valid. This means that $Sb(\frac{a}{x})P$ is true under every interpretation. In particular then, given any one interpretation, $Sb(\frac{a}{x})P$ is not only true under that interpretation, but also under every other interpretation exactly like that interpretation except that it assigns something different as the denotation of *a*. In other words, given any interpretation of $(x)P$, $Sb(\frac{a}{x})P$ is true under every *a*-variant of that interpretation. Therefore, given any interpretation of $(x)P$, $(x)P$ is true under that interpretation. Thus, $(x)P$ is valid.

The above principle is generally called the *principle of universal generalization.* We can use it to establish that universal generalizations are valid when we know that the scope of the quantifier already corresponds to a valid sentence.[9] This is particularly useful when taken together with the observation that all tautologies are valid. Given any tautology, we can construct a valid universal generalization by replacing individual constants by variables and appending universal quantifiers; for example, the sentence $[(Fa \equiv \sim Gb) \equiv \sim(Fa \equiv Gb)]$ is a tautology. From this, by universal generalization, we can conclude that $(y)[(Fa \equiv \sim Gy) \equiv \sim(Fa \equiv Gy)]$ is valid. Then by a second application of universal generalization, we can conclude that $(x)(y)[(Fx \equiv \sim Gy) \equiv \sim(Fx \equiv Gy)]$ is valid.

We can also construct a rule of universal generalization that will allow us to obtain new implications from old implications. This is the

[9] Here we say that the scope of the quantifier *corresponds* to a valid sentence rather than that it *is* a valid sentence. The scope of the quantifier will not be a sentence at all (unless the quantifier is a vacuous quantifier), because it contains a free variable, but we can construct a sentence that corresponds to it by substituting an individual constant for the free occurrences of the variables.

following: *Suppose P is any formula of the predicate calculus that contains free occurences of some variable, say "x," but not of any other variable, and suppose that "a" is some individual constant that does not occur in P. Then if some set of sentences, none of which contain a, implies* $Sb(\frac{a}{x})P$, *that set of sentences also implies* $(x)P$. The set of sentences $(x)(Fx \supset Gx)$, $(x)(Gx \supset Hx)$ implies the sentence $(Fa \supset Ha)$. Neither $(x)(Fx \supset Gx)$ nor $(x)(Gx \supset Hx)$ contains a. Therefore, this set of sentences also implies $(x)(Fx \supset Hx)$. Let us combine this principle with the previous principle and call the combined principle the *principle of universal generalization.*

Why is this principle true? Suppose we have a set of sentences, $Q_1, \ldots, Q_n$, none of which contains a, and suppose this set implies $Sb(\frac{a}{x})P$, where P does not contain a either. By definition, this means that $[(Q_1 \& \ldots \& Q_n) \supset Sb(\frac{a}{x})P]$ is valid. As $Q_1, \ldots, Q_n$ are all sentences, they contain no free occurrences of x, so

$$Sb(\tfrac{a}{x})[(Q_1 \& \ldots \& Q_n) \supset P]$$

is the same sentence as $[(Q_1 \& \ldots \& Q_n) \supset Sb(\frac{a}{x})P]$. Thus

$$Sb(\tfrac{a}{x})[(Q_1 \& \ldots \& Q_n) \supset P]$$

is also valid. Furthermore, as a does not occur in any of $Q_1, \ldots, Q_n$ or P, a does not occur in $[(Q_1 \& \ldots \& Q_n) \supset P]$. Thus, by the first principle of universal generalization, $(x)[(Q_1 \& \ldots \& Q_n) \supset P]$ is valid. x does not occur free in $(Q_1 \& \ldots \& Q_n)$, so by E4, $(x)[(Q_1 \& \ldots \& Q_n) \supset P]$ is equivalent to $[(Q_1 \& \ldots \& Q_n) \supset (x)P]$, and consequently this latter sentence is also valid. Therefore, the set of sentences $Q_1, \ldots, Q_n$ implies $(x)P$.

The use of this second principle of universal generalization corresponds to a frequently used pattern of informal reasoning. Suppose we want to prove that something is true of all objects in a certain domain, say the integers. In order to prove that it is true of all integers, we might reason as follows. First we pick some arbitrary integer, and show that it is true of that integer. Then we observe that our proof that it is true of that integer does not assume anything about that integer which is not true of all integers, and so the same proof will work for any other integer as well. Thus we conclude that it is true of every integer. Observing that the proof does not assume anything about the integer we chose that is not true of all integers is the same as observing that the premises of our proof do not refer to that integer—that they do not contain a. So we are in effect just using universal generalization.

III.12 EQUIVALENT FORMULAS

The concept of equivalence defined in Section III.10 only applies to sentences. Using that concept it makes no sense to talk about the equivalence of formulas that are not sentences. However, it is useful to extend this concept of equivalence so that we can talk about the equivalence of formulas in general.

First we need the concept of a closure of a formula. Given a formula of the predicate calculus, we say that a sentence is a *closure* of that formula if, and only if, the sentence results from appending universal quantifiers to the front of the formula to bind all free variables. For example, $(x)[(y)Fy \supset Fx]$ is a closure of $[(y)Fy \supset Fx]$; similarly, $(x)(y)(Fx \supset Fy)$ and $(y)(x)(Fx \supset Fy)$ are both closures of $(Fx \supset Fy)$. If more than one variable occur free in a formula, then there will be more than one closure of that formula, because the quantifiers can be appended in different orders. However, all of the closures will be equivalent to one another, because the order in which we write a string of universal quantifiers makes no difference. For example, $(x)(y)Fxy$ and $(y)(x)Fxy$ are equivalent, because $(x)(y)Fxy$ is true under an interpretation if, and only if, Fab is true under every b-variant of every a-variant; that is, if and only if Fab is true no matter what we assign as the denotations of a and b. But this is the same as saying that Fab is true under every a-variant of every b-variant, which is the condition under which $(y)(x)Fxy$ is true. So $(x)(y)Fxy$ and $(y)(x)Fxy$ are true under exactly the same conditions, and so their biconditional is valid, making them equivalent. This same argument works in general to show that it never makes any difference in which order we write a string of universal quantifiers.[10] Thus all closures of any given formula are equivalent.

Now we extend the notion of equivalence to formulas that are not sentences: two formulas are *equivalent* if, and only if, the two formulas contain free occurrences of exactly the same variables and the closures of their biconditional are valid. So $(x)Fxy$ and $\sim(\exists x) \sim Fxy$ are equivalent formulas, because $(y)[(x)Fxy \equiv \sim(\exists x) \sim Fxy]$ is valid. The closures of the biconditional of two formulas can be valid without the

[10] However, it does make a difference in what order we write a string of quantifiers containing both existential and universal quantifiers; for example, $(x)(\exists y)Fxy$ is not equivalent to $(\exists y)(x)Fxy$.

formulas containing free occurrences of the same variables; for instance, the closures of $[(Fx \vee \sim Fx) \equiv (Fy \vee \sim Fy)]$ are valid. But if two formulas do contain free occurrences of exactly the same variables, then there is a natural intuitive significance to saying that the closures of their biconditional are valid—that they are equivalent. Recall that formulas that are not sentences symbolize the forms of relations. Thus if two such formulas are equivalent, then they must always be the symbolizations of relations that have the same extensions. The formulas $(Fx \& Gy)$ and $\sim(Gy \supset \sim Fx)$ are equivalent; therefore, $(x)(y)[(Fx \& Gy) \equiv \sim(Gy \supset \sim Fx)]$ is valid. Therefore, given any two things, x and y, they stand in the relation $(Fx \& Gy)$ to one another if, and only if, they stand in the relation $\sim(Gy \supset \sim Fx)$ to one another. In other words, the relations symbolized by $(Fx \& Gy)$ and $\sim(Gy \supset \sim Fx)$ have the same extensions.

As before, the significance of equivalence is that if two formulas are equivalent, then the result of replacing one by the other in any sentence yields a sentence that is equivalent to the original. $(Fx \equiv \sim Gy)$ and $\sim(Fx \equiv Gy)$ are equivalent; therefore, $(x)(\exists y)[Hxy \equiv (Fx \equiv \sim Gy)]$ is equivalent to $(x)(\exists y)[Hxy \equiv \sim(Fx \equiv Gy)]$. The reason for this is simply that if two formulas are equivalent, then (as we saw above) they correspond to relations having the same extensions. It is only the extension of a relation that is relevant in determining the truth value of a sentence of the predicate calculus, so two sentences that are identical except that one contains an occurrence of the one formula where the other contains an occurrence of the other formula will always have the same truth value. Therefore, they will be equivalent.[11]

[11] The above argument is not as rigorous as might be desired. To make it more rigorous one must proceed as follows. First it must be shown that given any two formulas, P and Q, if they are equivalent then $\sim P$ and $\sim Q$ are equivalent, and given any variable, say x, $(x)P$ and $(x)Q$ are equivalent, and $(\exists x)P$ and $(\exists x)Q$ are equivalent. Next it must be shown that if P and Q are equivalent, then given any third formula R, $(P \& R)$ is equivalent to $(Q \& R)$, $(R \& P)$ is equivalent to $(R \& Q)$, $(P \vee R)$ is equivalent to $(Q \vee R)$, $(R \vee P)$ is equivalent to $(R \vee Q)$, $(P \supset R)$ is equivalent to $(Q \supset R)$, $(R \supset P)$ is equivalent to $(R \supset Q)$, $(P \equiv R)$ is equivalent to $(Q \equiv R)$, and $(R \equiv P)$ is equivalent to $(R \equiv Q)$. This is not difficult to show. Once this has been shown, it follows that the result of substituting P for Q in any formula yields a formula that is equivalent to the original. This is because we can simply build up the two larger formulas out of P and Q using our rules for constructing formulas of the predicate calculus, and using the above results, it can be concluded that at each step of the construction the two formulas at that point are equivalent because the formulas immediately before were equivalent.

This means we can construct a rule of replacement for the predicate calculus analogous to the rule of tautological replacement for the propositional calculus: if one sentence can be obtained from a second by replacing one of the parts of the first by a formula to which it is equivalent, then the second sentence is equivalent to the first sentence. This is a generalized version of the rule discussed in Section III.10.

This rule of replacement can be generalized still further. As stated, it only talks about replacement in sentences, but we can talk about replacement in formulas in general: *if one formula can be obtained from a second by replacing one of the parts of the first by a formula to which it is equivalent, then the second formula is equivalent to the first formula.* To show that this rule holds, suppose that P is a formula containing free occurrences of some variables, say $x_1, \ldots, x_n$, and Q is a formula obtained from P by replacing some part of P by a formula to which it is equivalent. Now take some individual constants, say $a_1, \ldots, a_n$, which do not occur in either P or Q, and substitute them for the free occurrences of variables in P and Q, obtaining $Sb\binom{a_1 \cdots a_n}{x_1 \ldots x_n}P$ and $Sb\binom{a_1 \cdots a_n}{x_1 \ldots x_n}Q$. These latter formulas are sentences, and $Sb\binom{a_1 \cdots a_n}{x_1 \ldots x_n}Q$ can be obtained by replacement from $Sb\binom{a_1 \cdots a_n}{x_1 \ldots x_n}P$ in the same way that Q was obtained from P. Thus, by our above rule of replacement in sentences, $Sb\binom{a_1 \cdots a_n}{x_1 \ldots x_n}Q$ is equivalent to $Sb\binom{a_1 \cdots a_n}{x_1 \ldots x_n}P$; that is, the biconditional $[Sb\binom{a_1 \cdots a_n}{x_1 \ldots x_n}Q \equiv Sb\binom{a_1 \cdots a_n}{x_1 \ldots x_n}P]$ is valid. This is the same sentence as $Sb\binom{a_1 \cdots a_n}{x_1 \ldots x_n}(Q \equiv P)$, so this latter sentence is valid. But then, by universal generalization, $(x_1)(x_2)\ldots(x_n)(Q \equiv P)$ is valid; that is, P and Q are equivalent formulas. Therefore, the strengthened rule of replacement holds.

III.12 EXERCISES

For each of the following pairs of formulas, tell whether they can be obtained from one another by a single application of the rule of replacement using one of the equivalences E1–E27. If they can, which equivalence is used?

1. $(x)(Fx \supset Gx)$ / $(x)({\sim}Fx \vee Gx)$
2. $(x)(y)Fxy$ / $(z)(w)Fzw$
3. $(x)[Wx \equiv ((\exists y)Fy \supset Hx)]$ / $(x)[Wx \equiv (y)(Fy \supset Hx)]$
4. ${\sim}(x)(Fx \equiv Gx)$ / $(x)(Fx \equiv {\sim}Gx)$
5. $[P \mathbin{\&} (x)(Fx \supset Hx)]$ / $[P \mathbin{\&} ((\exists x)Fx \supset Hx)]$
6. $(x) \sim (Fx \mathbin{\&} Gxy)$ / $(x)({\sim}Fx \mathbin{\&} {\sim}Gxy)$

III.13 SUBSTITUTION

In Section III.10, a number of implications and equivalences were listed, and it was shown that they could be regarded as implication and equivalence *schemes*—any sets of formulas that result from making appropriate substitutions in these implications and equivalences are also implications and equivalences. This can be generalized by defining a general concept of a substitution instance. Just as in the propositional calculus, it can be shown that in the predicate calculus there are certain kinds of substitutions that can be performed which, if begun with a valid sentence, will always lead to another valid sentence.

The notion of a substitution instance is somewhat more complicated for the predicate calculus than it was for the propositional calculus. In the propositional calculus all formulas were constructed out of one kind of building block—sentential letters. But in the predicate calculus there are three different kinds of building blocks—relation symbols, individual constants, and variables—and in constructing substitution instances we can substitute things for each of these basic elements. This leads to three different kinds of substitutions. In each case we will see that a sentence obtained by substitution from a valid sentence is valid.

a. Substitution for Individual Constants. The first kind of substitution that we can perform is the substitution of one individual constant for every occurrence of another individual constant; for example, we can substitute a for every occurrence of b in the formula

$$[Hb \supset ((x)Gxb \supset \sim P)]$$

giving us $[Ha \supset ((x)Bxa \supset \sim P)]$.

It is easy to see that if a sentence is obtained from a valid sentence by substitution for individual constants, then it will also be valid. This is because the *valid* sentence must be true no matter what we choose for the denotation of the individual constants. Thus it must remain true when we substitute the new individual constants for the original ones. This kind of substitution always leads from valid sentences to valid sentences.

b. Substitution for Bound Variables. The second kind of substitution is that of substituting one variable for every occurrence of another variable that is bound by a particular quantifier, together with substitution for the variable in the quantifier itself, provided that the new variable does not occur in the scope of the quantifier in the original formula. We can substitute y for the occurrences of x bound by the first quantifier in $(x)[Hx \ \& \ (Gx \supset (\exists x) \sim Hx)]$ to obtain

$$(y)[Hy \ \& \ (Gy \supset (\exists x) \sim Hx)]$$

Notice that we must replace the quantifier "(x)" by "(y)," and that we leave the occurrence of x bound by the existential quantifier unchanged. Similarly, we might substitute z for every occurrence of x bound by the second quantifier in $[(x)(Fx \supset Gx) \ \& \ (x)(Gx \supset Hx)]$ to obtain the formula $[(x)(Fx \supset Gx) \ \& \ (z)(Gz \supset Hz)]$. But we cannot substitute x for y in $(x)(y)(Fx \supset Fy)$ to obtain $(x)(x)(Fx \supset Fx)$, because x already occurs in $(Fx \supset Fy)$, which is the scope of "(y)."

Now it must be shown that this kind of substitution always leads from valid sentences to valid sentences. Suppose we begin with a sentence Q containing the formula $(x)P$ as one of its parts. Then we substitute another variable, say y, for x, obtaining the sentence R, which is exactly like Q except that it contains $(y)Sb(\frac{y}{x})P$ in place of $(x)P$. If y does not occur in P, then by E26, $(y)Sb(\frac{y}{x})P$ is equivalent to $(x)P$. Then by replacement, R is equivalent to Q. Therefore, if Q is valid, so is R. Consequently, if we begin with a valid sentence which contains $(x)P$ as one of its parts, and then substitute $(y)Sb(\frac{y}{x})P$ for $(x)P$, the new sentence will also be valid. The analogous thing is true for existential quantifiers, and so this kind of substitution will always lead from valid sentences to valid sentences.

The reason we must make the condition that y does not occur in P is if y does occur in P, then $(x)P$ and $(y)Sb(\frac{y}{x})P$ may not be equivalent. If for example, P were the formula Hxy, then $(x)Hxy$ says that everything stands in the relation H to y. But $(y)Hyy$, which would be $(y)Sb(\frac{y}{x})P$, says that everything stands in the relation H to itself, which is something quite different. Thus $(x)P$ and $(y)Sb(\frac{y}{x})P$ may not be equivalent if y occurs in P, and so we cannot substitute the one for the other in a valid sentence and be sure that we will get another sentence that is valid. For instance,

$$[((\exists x)Fx \ \& \ (\exists y) \sim Fy) \supset (\exists x)(\exists y)(Fx \ \& \ \sim Fy)]$$

is valid, but if we substitute x for the occurrences of y in the consequent, we get $[((\exists x)Fx \ \& \ (\exists y) \sim Fy) \supset (\exists x)(\exists x)(Fx \ \& \ \sim Fx)]$. Clearly the consequent of the latter conditional can never be true, so that conditional is true if, and only if, the antecedent is false. But it is easy to construct interpretations under which the antecedent is true. We might let the domain be the set of all integers and let F mean "is even." Under this interpretation the antecedent is true, and so the conditional is false, hence it is not valid.

c. Substitution for Relation Symbols. Given a formula containing a relation symbol, under certain circumstances we can substitute another formula for this relation symbol each time it occurs in the original formula. The simplest case occurs when the relation symbol is a sentential letter. Then we can substitute any other sentence for it as long as we make the same substitution for each occurrence of that sentential letter in the formula. This is just like substitution in the propositional calculus. And, for the same reasons as in the propositional calculus, if we perform such a substitution in a valid sentence the resulting sentence will also be valid.

If the relation symbol is not a sentential letter the situation becomes more complicated. Let us look at an example of this. The sentence $\sim(x)(y)(Hxy \;\&\; \sim Hxx)$ is valid. To see that it is, suppose instead that $(x)(y)(Hxy \;\&\; \sim Hxx)$ were true under some interpretation. This says that given any things, x and y, x stands in the relation H to y but does not stand in the relation H to itself. But if this is true for any things x and y, it is also true if x and y are the same thing. Thus letting x be y, we would have to have $(y)(Hyy \;\&\; \sim Hyy)$ true, which is impossible. Thus $(x)(y)(Hxy \;\&\; \sim Hxx)$ cannot be true under any interpretation, and so $\sim(x)(y)(Hxy \;\&\; \sim Hxx)$ is true under every interpretation; that is, it is valid. All of this is true regardless of what relation is symbolized by H. Thus we might replace H throughout by a formula that symbolizes a more complicated relation, and the resulting sentence would still be valid; for example, we might replace H by the relation $(\exists z)R(1)z(2)$. Hxy says that x stands in the relation H to y, so this would then be symbolized as $(\exists z)Rxzy$. Hxx says that x stands in the relation H to itself, so this would be symbolized as $(\exists z)Rxzx$. Thus replacing H by the relation $(\exists z)R(1)z(2)$ in the sentence

$$\sim(x)(y)\;(Hxy \;\&\; \sim Hxx)$$

would give us the sentence $\sim(x)(y)[(\exists z)Rxzy \;\&\; \sim(\exists z)Rxzx]$. This sentence will also be valid, for the same reasons that the original sentence was valid.

The expression $(\exists z)R(1)z(2)$ is not a formula of the predicate calculus because of the numbers in parentheses, so it is not correct to say that we are substituting a formula for each occurrence of H. What we are doing is substituting an expression that symbolizes another relation. The reason we must use these expressions with numbers is that we must substitute the same relation for each occurrence of H, and if the different occurrences of H are followed by different sequences of individual symbols, then we must insert those individual symbols into $(\exists z)R(1)z(2)$ in the proper order; the numbers tell us in what order they go. In making the substitution for Hxy, x is the first individual symbol following H, and y is the second, so x is written in place of (1), and y is written in place of (2) to give us the formula $(\exists z)Rxzy$. On the other hand, in making the substitution for Hxx, x occupies both the first and second places after H, so x is written in place of both (1) and (2), giving us $(\exists z)Rxzx$.

In general then, if we start with a valid sentence and replace relation symbols by other relations, the resulting sentence will still be valid. The reason for this is that if a sentence containing a relation symbol is valid, then it will be true regardless of how the relation symbol is interpreted; that is, regardless of what relation is assigned to the relation symbol. In particular, if we assign any relation that can be symbolized by the more complex expression we substitute for the relation symbol in constructing the substitution instance, the sentence will be true. But to say that the sentence will be true regardless of which of those relations are assigned to the relation symbol means that the substitution instance of the sentence is valid. Let us look at some examples of this kind of substitution. Suppose we begin with the sentence $[(x)Fxa \supset Faa]$. Then we might substitute the relation $[H(1)(2) \supset (\exists y)(z)(G(1)yz(2) \ \& \ {\sim}H(2)z)]$ for F. Making this substitution for Fxa gives us $[Hxa \supset (\exists y)(z)(Gxyza \ \& \ {\sim}Haz)]$, and making this substitution for Faa gives us $[Haa \supset (\exists y)(z)(Gayza \ \& \ {\sim}Haz)]$. Thus we get the sentence $[(x)[Hxa \supset (\exists y)(z)(Gxyza \ \& \ {\sim}Haz)] \supset [Haa \supset (\exists y)(z)(Gayza \ \& \ {\sim}Haz)]]$. To take another example, suppose we begin with the sentence $(x)(y)(Ryx \supset Ryx)$ and substitute

$$(\exists z)(I(1)z \ \& \ Iz(2))$$

for R. Then we get the sentence

$$(x)(y)[(\exists z)(Ixz \ \& \ Izy) \supset (\exists z)(Iyz \ \& \ Izx)]$$

The three kinds of substitutions that can be performed in the predicate calculus correspond to the three kinds of building blocks out of which formulas of the predicate calculus are constructed, and for each kind of substitution, if we start with a valid sentence and then perform the substitution the resulting sentence will also be valid. We say that one sentence of the predicate calculus is a *substitution instance* of a second sentence if, and only if, the first can be obtained from the second by making substitutions of the above three kinds. Then what we have shown is that every substitution instance of a valid sentence is valid.

Just as in the propositional calculus, we also talk about one *set* of sentences being a substitution instance of a second *set* of sentences when the first results from making the same substitutions in every sentence in the second set. So, the set of sentences $(x)Fxa$, $(\exists y)(Gya \supset Fya)$ is a substitution instance of the set of sentences $(x)Px$, $(\exists z)(Gza \supset Pz)$. Here we have substituted y for z and $F(1)a$ for P.

Substitution gives us a way of obtaining valid sentences from valid sentences. It also gives us a way of obtaining implications from implications. If a set of sentences, $P_1, \ldots, P_n$, implies a sentence Q, then if the same substitutions are performed in $P_1, \ldots, P_n$ and Q, the resulting substitution instances of $P_1, \ldots, P_n$ will imply the substitution instance of Q. To see that this is true, suppose $P_1, \ldots, P_n \square= Q$. Then $[(P_1 \& \ldots \& P_n) \supset Q]$ is valid. If $R_1, \ldots, R_n$ are the substitution instances of $P_1, \ldots, P_n$, respectively, and S is the corresponding substitution instance of Q, then $[(R_1 \& \ldots \& R_n) \supset S]$ is a substitution instance of $[(P_1 \& \ldots \& P_n) \supset Q]$, so it is valid. Consequently, $R_1, \ldots, R_n \square= S$.

The concept of a substitution instance defined above applies only to sentences, not formulas in general; that is, we have only discussed the case of one *sentence* being a substitution instance of another *sentence*. However, for certain purposes it is useful to have the general concept of a formula being a substitution instance of a formula. With one slight change, the three kinds of substitutions that can be performed are the same as before. The change concerns substitution for relation symbols. In substituting for the relation symbols in a sentence to obtain another sentence, we cannot introduce any new free occurrences of variables in the expressions that are substituted for the relation symbols. If we did, the resulting formula would not be a sentence. If we were to substitute $\sim Rxz(1)$ for F in $(y)Fy$ we would get $(y) \sim Rxzy$, which contains free occurrences of x and z. However, in our general concept of a substitution instance there is no reason to disallow this, provided only that the occurrences of variables that are free in the expressions we substitute for the relation symbols are also free in the whole formula. This means, for example, that we cannot substitute $\sim Rxz(1)$ for F in

$(x)Fx$ to get $(x) \sim Rxzx$ because although x occurs free in $\sim Rxz(1)$, it does not occur free in $(x) \sim Rxzx$.

We must bear in mind that on this generalized concept of a substitution instance, a substitution instance of a sentence need not be a sentence. It makes no sense to talk about the validity of a formula that is not a sentence, so the rule that a substitution instance of a valid sentence is valid must be qualified to read "a sentence that is a substitution instance of a valid sentence is valid."

The reason we are interested in talking about substitution for formulas that are not sentences is that this gives us a way of generating new equivalences. *If P and Q are any formulas of the predicate calculus, and P is equivalent to Q, and R and S are formulas that result from making the same substitutions in P and Q, respectively, and R and S contain free occurrences of the same variables, then R is equivalent to S.* This is proven as follows. Suppose P and Q are equivalent, and $x_1, \ldots, x_n$ are the variables having free occurrences in P and Q. Then $(x_1)(x_2)\ldots(x_n)(P \equiv Q)$ is valid. R and S may contain free occurrences of variables other than $x_1, \ldots, x_n$, because such free occurrences can be introduced in the process of substituting for relation symbols. Suppose the additional variables are $y_1, \ldots, y_m$. Next choose individual constants $a_1, \ldots, a_m$ that do not occur in R or S. Then $Sb\binom{a_1 \ldots a_m}{y_1 \ldots y_m}R$ is also a substitution instance of P, and $Sb\binom{a_1 \ldots a_m}{y_1 \ldots y_m}S$ is also a substitution instance of Q, and furthermore these new formulas contain free occurrences only of the same variables as P and Q. Therefore,

$$(x_1)\ldots(x_n)[Sb\binom{a_1 \ldots a_m}{y_1 \ldots y_m}R \equiv Sb\binom{a_1 \ldots a_m}{y_1 \ldots y_m}S]$$

is a substitution instance of $(x_1)\ldots(x_n)(P \equiv Q)$. The latter is valid, thus so must the former be valid. But then, by universal generalization, $(y_1)\ldots(y_m)(x_1)\ldots(x_n)(R \equiv S)$ must also be valid. Hence R is equivalent to S.

This means that we can interpret the equivalence schemes E1–E27 as applying to formulas that are not sentences as well as to sentences, because a substitution instance of a sentence need not be a sentence. By E24 for example, we can conclude that the formulas $[Fy \supset (x)Gx]$ and $(x)[Fy \supset Gx]$ are equivalent.

Substitution gives us one way to obtain new valid sentences from old which is particularly useful in connection with tautologies. Every formula of the propositional calculus is a sentence of the predicate calculus, and every tautology of the propositional calculus is a valid sentence of the predicate calculus. An interpretation of a sentence of the predicate calculus that is also a formula of the propositional calculus does nothing more than assign truth values to its atomic parts. Thus to say that such a sentence is valid (true under every interpretation) is just to say that it is true no matter what truth values we assign to its atomic parts; that is, it is a tautology. Thus our previous discussion of tautologies gives us a large stock of valid sentences with which to work. Notice further that a substitution instance of a sentence that is a formula of the propositional calculus need not itself be a formula of the propositional calculus; for example $[(x)Fx \vee \sim(x)Fx]$ is a substitution instance of $(P \vee \sim P)$. And we know that $(P \vee \sim P)$ is a tautology, so we can conclude immediately that $[(x)Fx \vee \sim(x)Fx]$ is valid.

This suggests that, in the predicate calculus, we should define the notion of a tautology more broadly. Let us say that a sentence of the predicate calculus is a tautology if, and only if, it is either a tautologous formula of the propositional calculus, or it can be obtained by substitution from a tautologous formula of the propositional calculus. Thus, $[(x)Fx \vee \sim(x)Fx]$ is a tautology. Tautologies will give us a simple core of valid sentences with which to work in order to establish that other sentences are valid.

Corresponding to the extended concept of a tautology we can define extended concepts of tautological implication and tautological equivalence. If $P_1, \ldots, P_n$ and Q are any sentences of the predicate calculus, we say that $P_1, \ldots, P_n$ *tautologically implies* Q if, and only if,

$$[(P_1 \,\&\, \ldots \,\&\, P_n) \supset Q]$$

is a tautology. For example, $(x)Fx$ and $(x)Gx$ tautologically implies $[(x)Fx \,\&\, (x)Gx]$. Analogously, if P and Q are any formulas of the predicate calculus, we say that P and Q are *tautologically equivalent* if, and only if, $(P \equiv Q)$ is a tautology.

III.13 EXERCISES

A. Construct the following substitution instances:

1. Substitute a for b and z for bound occurrences of x in $(Fa \vee (\exists x)Hbx)$.
2. Substitute $(Fa(1) \vee G(2))$ for H in $(x)(y)(Hxy \supset Hyx)$.
3. Substitute $(Fa(1) \vee G(2))$ for H in $[Hab \equiv (\exists x)(y)(\sim Hxy \supset Hax)]$.
4. Substitute $(\exists y)H(1)yb$ for F and b for a in $(Fa \,\&\, \sim Fb)$.

B. What substitutions might be performed in order to show that the formulas after the slashes are substitution instances of those before the slashes?

1. $Fa \,/\, (\exists x)Haxb$
2. $(Fab \supset (\exists x)Fax) \,/\, (Gbac \supset (\exists z)Gzac)$
3. $[P \equiv (Fa \vee \sim P)] \,/\, [(x)(\exists y)Hxy \equiv ((\exists z)Gzac \vee \sim(x)(\exists y)Hxy)]$
4. $(x)(\exists y)(Hxy \equiv \sim(Hay \supset Hxa)) \,/\, (z)(\exists w)\{(Mwaz \vee Nzw) \equiv \sim[(Mwaa \vee Naw) \supset (Maaz \vee Nza)]\}$

C. 1. Give an example of a sentence that is valid but not a tautology.
2*. Show that every substitution instance of an inconsistent sentence is inconsistent.

III.14 DERIVATIONS

In the propositional calculus we discovered a number of rules that helped us to show that formulas were tautologies or that formulas were tautologically implied by other formulas. Then we put those rules together and defined the notion of a derivation in the propositional calculus. The same thing will now be done for the predicate calculus.

A derivation in the predicate calculus, or *derivation* for short, will have the same structure as a derivation in the propositional calculus. It consists of a sequence of consecutively numbered lines, and each line begins with premise numbers, then the line number, then the sentence written on that line, and finally an explanation of how we got that line. The only difference will be that we will use an expanded list of rules of inference. The rules of inference for the predicate calculus are the following:

RULE P: PREMISE INTRODUCTION: On any line of a derivation, we can write any sentence we want provided we let the premise number be the line number of that line.

RULE I: IMPLICATION: If, according to one of the implication schemes I1–I17, a set of sentences appearing on one or more lines of a derivation implies a further sentence, then we can write that further sentence on any later line of the derivation, taking as premise numbers all of the premise numbers of the set of sentences.

RULE C: CONDITIONALIZATION: If some sentence, *Q*, appears on a line of a derivation, and another sentence, *P*, is among the premises of Q, then on any later line of the derivation we can write the conditional $(P \supset Q)$, taking for premise numbers all of the premise numbers of Q except for the line number of P.

RULE R: REPLACEMENT: If one sentence appears on a line of a derivation, and a second sentence can be obtained from the first by replacing some part of it by another formula to which it is equivalent by one of the equivalences E1–E27, then we can write the second sentence on any later line of the derivation, taking for premise numbers all of the premise numbers of the first sentence.

RULE UG: UNIVERSAL GENERALIZATION: If P is any formula of the predicate calculus, which contains free occurrences of some variable, say x, but not of any other variable, and a is some individual constant that does not occur in P, then if $Sb(\frac{a}{x})P$ appears on any line of a derivation, and it either has no premises or none of its premises contain a, then we can write $(x)P$ on any later line of the derivation, taking for premise numbers all of the premise numbers of $Sb(\frac{a}{x})P$.

The first four rules of inference for the predicate calculus are essentially the same as the corresponding rules for the propositional calculus. The last is based on the principle of universal generalization. In the light of our previous discussion of these rules, it should be clear that *every line of a derivation is implied by its premises if it has any and is valid if it does not have any*. A sentence that can be obtained on a line of a derivation without any premise numbers is a *theorem of the predicate calculus*. Therefore, all theorems of the predicate calculus are valid; that is, derivations in the predicate calculus are sound. It will be shown in Section III.16 that the converse is also true, namely, all valid sentences are theorems of the predicate calculus—derivations in the predicate calculus are complete.

If the last line of a derivation consists of a sentence without any premise numbers, then that derivation is said to be a derivation *of* that sentence. And if the last line of a derivation consists of a sentence with premise numbers, then that derivation is said to be a derivation *of* that sentence *from* those premises.

Now let us look at some sample derivations without, for the moment, worrying about the strategies involved in constructing them. Suppose we want to construct a derivation of $(x)[Fx \supset (Fx \vee Gx)]$. We might proceed as follows:

(1)	1.	Fa	P
(1)	2.	$(Fa \vee Ga)$	I, I (I3)

	3.	$[Fa \supset (Fa \vee Ga)]$	C, 1, 2
	4.	$(x)[Fx \supset (Fx \vee Gx)]$	UG, 3

To take another example, suppose we want to derive $(x)(\exists y)Fxy$ from $(\exists y)(x)Fxy$. We might proceed as follows:

(1)	1.	$(\exists y)(x)Fxy$	P
(2)	2.	$(x)Fxa$	P
(2)	3.	Fba	I, 2 (I16)
(2)	4.	$(\exists y)Fby$	I, 3 (I17)
(2)	5.	$(x)(\exists y)Fxy$	UG, 4
	6.	$[(x)Fxa \supset (x)(\exists y)Fxy]$	C, 2, 5
	7.	$(y)[(x)Fxy \supset (x)(\exists y)Fxy]$	UG, 6
	8.	$[(\exists y)(x)Fxy \supset (x)(\exists y)Fxy]$	R, 7 (E25)
(1)	9.	$(x)(\exists y)Fxy$	I, 1, 8 (I9)

Suppose next we want to derive $(\exists y)Gy$ from $(x)(Fx \supset Fy)$ and $(\exists z)Fz$. We might proceed as follows:

(1)	1.	$(x)(Fx \supset Gx)$	P
(2)	2.	$(\exists z)Fz$	P
(3)	3.	Fa	P
(1)	4.	$(Fa \supset Ga)$	I, 1 (I16)
(1, 3)	5.	Ga	I, 3, 4 (I9)
(1, 3)	6.	$(\exists y)Gy$	I, 5 (I17)
(1)	7.	$[Fa \supset (\exists y)Gy]$	C, 3, 6
(1)	8.	$(z)[Fz \supset (\exists y)Gy]$	UG, 7
(1)	9.	$[(\exists z)Fz \supset (\exists y)Gy]$	R, 8 (E25)
(1, 2)	10.	$(\exists y)Gy$	I, 2, 9 (I9)

There are definite strategies that are often useful in constructing derivations, just as there were in the case of PC derivations. The strategies for conjunctions, disjunctions, conditionals, and biconditionals are the same as before. The strategy for negations must be augmented by the observation that, by E22, $\sim(x)P$ is equivalent to $(\exists x) \sim P$, and $\sim(\exists x)P$ is equivalent to $(x) \sim P$. There are also strategies now for universal generalizations and existential generalizations.

A Universal Generalizations

The most common strategy to derive something from a universal generalization is to use Rule I with I16 to obtain a sentence without the universal quantifier. This is called *universal specification.* Suppose we want to derive $(x)(\sim Gx \supset \sim Fx)$ from $(x)(Fx \supset Gx)$. Then we might proceed as follows:

(1)	1.	$(x)(Fx \supset Gx)$	P
(1)	2.	$(Fa \supset Ga)$	I, 1 (I16) (universal specification)
(1)	3.	$(\sim Ga \supset \sim Fa)$	R, 2 (E16)
(1)	4.	$(x)(\sim Gx \supset \sim Fx)$	UG, 3

Conversely, suppose we want to derive a universal generalization from something else. A common strategy is to derive it without the universal quantifier, and then use UG, which is what was done in the above derivation.

B Existential Generalizations

We can frequently derive an existential generalization from something else by deriving the sentence without the quantifier and then using I17. This is called *existential generalization.* For example, suppose we want to derive $(\exists x)Gx$ from $(x)(\sim Gx \supset Gx)$. We might proceed as follows:

(1)	1.	$(x)(\sim Gx \supset Gx)$	P
(1)	2.	$(\sim Ga \supset Ga)$	I, 1 (I16) (universal specification)
(1)	3.	$(\sim\sim Ga \vee Ga)$	R, 2 (E4)
(1)	4.	$(Ga \vee Ga)$	R, 3 (E1)
(1)	5.	Ga	R, 4 (E18)
(1)	6.	$(\exists x)Gx$	I, 5 (I17) (existential generalization)

On the other hand if we want to derive something from an existential generalization, the strategy is more complicated. Before considering the general strategy, let us look at an example of it. To derive $(\exists x)Gx$ from $(\exists x)(Fx \,\&\, Gx)$. We might proceed as follows:

(1)	1.	$(\exists x)(Fx \,\&\, Gx)$	P	
(2)	2.	$(Fa \,\&\, Ga)$	P	
(2)	3.	Ga	I, 2 (I2)	
(2)	4.	$(\exists x)Gx$	I, 3 (I17)	existential specification
	5.	$[(Fa \,\&\, Ga) \supset (\exists x)Gx]$	C, 2, 4	
	6.	$(x)[(Fx \,\&\, Gx) \supset (\exists x)Gx]$	UG, 5	
	7.	$[(\exists x)(Fx \,\&\, Gx) \supset (\exists x)Gx]$	R, 6 (E32)	
(1)	8.	$(\exists x)Gx$	I, 1, 7 (I9)	

This strategy is called *existential specification.* In order to derive something from $(\exists x)(Fx \mathbin{\&} Gx)$, we took as a new premise $(Fa \mathbin{\&} Ga)$, which is $Sb(^{a}_{x})(Fx \mathbin{\&} Gx)$. Then we proceeded to derive our conclusion $(\exists x)Gx$ from this new premise. Next we conditionalized, obtaining

$$[(Fa \mathbin{\&} Ga) \supset (\exists x)Gx]$$

Then we used UG, to obtain $(x)[(Fx \mathbin{\&} Gx) \supset (\exists x)Gx]$. Because x does not occur free in $(\exists x)Gx$, we were able to use Rule R with E25 to turn the universal generalization into $[(\exists x)(Fx \mathbin{\&} Gx) \supset (\exists x)Gx]$, and finally, by modus ponens, we had our desired conclusion following from the desired premises.

Generalizing this strategy, suppose we want to derive some sentence Q from the premise $(\exists x)P$ together with some other premises (whose numbers are indicated in the following derivation by the dots in the left-hand column). We choose an individual constant a, which does not occur in Q or P or any of the premises, and proceed:

(1)	1.		P	premises other than $(\exists x)P$	
(2)	2.		P		
	⋮				
(n)	n.	$(\exists x)P$	P		
$(n+1)$	$n+1$.	$Sb(^{a}_{x})P$	P	derivation of Q from $Sb(^{a}_{x})P$	
	⋮				
$(n+1, \ldots)$	k.	Q			existential specification
...	$k+1$.	$[Sb(^{a}_{x})P \supset Q]$	C, $(n+1)$, k		
...	$k+2$.	$(x)(P \supset Q)$	UG, $k+1$		
...	$k+3$.	$[(\exists x)P \supset Q]$	R, $k+2$ (E25)		
$(n, \ldots)$	$k+4$.	Q	I, n, $k+3$ (I9)		

Notice that the second and third derivation on page 151 both use existential specification.

The intuitive rationale for existential specification is the following. Suppose we know that $(\exists x)Fx$ is true, that something is an F. From this we want to derive the conclusion Q. We might begin our argument by saying, "We know that something is an F; call it 'a.'" This is the same as taking Fa as a new premise. If we can then derive Q from Fa without assuming anything about a, then Q will follow from Fa regardless of what a is, and so Q will follow from the mere fact that there is something that is an F. This reasoning corresponds to the last three steps of existential specification. Notice that in order for us to be able to use UG on line $(k + 2)$, a cannot occur in P or Q or any of the premises. This corresponds in the intuitive reasoning pattern to requiring that we have not assumed anything about a. It is important to keep this restriction in mind, because it is a very common mistake to overlook it.

These strategies combine to give us one basic strategy for derivations in the predicate calculus: *eliminate quantifiers from the premises using existential and universal specification, then proceed as in the propositional calculus to derive the conclusion without quantifiers, and finally replace quantifiers using existential and universal generalization.* Let us look at some examples of this.

Suppose now we want to derive $(\exists x)(z)(\exists y)Fxyz$ from $(z)(\exists y)(z)Fxyz$. We might proceed as follows:

(1)	1.	$(x)(\exists y)(z)Fxyz$	P
(1)	2.	$(\exists y)(z)Fayz$	I, 1 (I16) (universal specification)
(3)	3.	$(z)Fabz$	P (beginning existential specification)
(3)	4.	$Fabc$	I, 3 (I16) (universal specification)
(3)	5.	$(\exists y)Fayc$	I, 4 (I17) (existential generalization)
(3)	6.	$(z)(\exists y)Fayz$	UG, 5
(3)	7.	$(\exists x)(z)(\exists y)Fxyz$	I, 6 (I17) (existential generalization)
	8.	$[(z)Fabz \supset (\exists x)(z)(\exists y)Fxyz]$	C, 3, 7
	9.	$(y)[(z)Fayz \supset (\exists x)(z)(\exists y)Fxyz]$	UG, 8
	10.	$[(\exists y)(z)Fayz \supset (\exists x)(z)(\exists y)Fxyz]$	R, 9 (E25)
(1)	11.	$(\exists x)(z)(\exists y)Fayz$	I, 2, 10 (I9) (completing existential specification)

Here we began by eliminating quantifiers in steps 1–4. This immediately gave us our conclusion without quantifiers. Then in steps 5–11 we replaced the quantifiers to give us our conclusion with the desired quantifiers.

To take a more difficult example, suppose we want to derive $(x)(Fx \supset {\sim}Px)$ from the set of premises $(x)[Gx \supset (y)(Fy \supset Hxy)]$, $(x)[Gx \supset (z)(Pz \supset {\sim}Hxz)]$, $(\exists x)Gx$:

(1)	1.	$(x)[Gx \supset (y)(Fy \supset Hxy)]$	P
(2)	2.	$(x)[Gx \supset (z)(Pz \supset {\sim}Hxz)]$	P

(3)	3.	$(\exists x)Gx$	P
(4)	4.	Ga	P (beginning existential specification)
(1)	5.	$[Ga \supset (y)(Fy \supset Hay)]$	I, 1 (I16)
(2)	6.	$[Ga \supset (z)(Pz \supset \sim Haz)]$	I, 2 (I16)
(1, 4)	7.	$(y)(Fy \supset Hay)$	I, 4, 5 (I9)
(2, 4)	8.	$(z)(Pz \supset \sim Haz)$	I, 4, 6 (I9)
(1, 4)	9.	$(Fb \supset Hab)$	I, 7 (I16)
(2, 4)	10.	$(Pb \supset \sim Hab)$	I, 8 (I16)
(2, 4)	11.	$(\sim \sim Pb \supset \sim Hab)$	R, 10 (E1)
(2, 4)	12.	$(Hab \supset \sim Pb)$	R, 11 (E16)
(1, 2, 4)	13.	$(Fb \supset \sim Pb)$	I, 9, 12 (I12)
(1, 2, 4)	14.	$(x)(Fx \supset \sim Px)$	UG, 13
(1, 2)	15.	$[Ga \supset (x)(Fx \supset \sim Px)]$	C, 4, 14
(1, 2)	16.	$(x)[Gx \supset (x)(Fx \supset \sim Px)]$	UG, 15
(1, 2)	17.	$[(\exists x)Gx \supset (x)(Fx \supset \sim Px)]$	R, 16 (E25)
(1, 2, 3)	18.	$(x)(Fx \supset \sim Px)$	I, 3, 17 (I9) (completing existential specification)

In this derivation, steps 1–10 involve eliminating the quantifiers from the premises. Steps 11–13 involve getting the desired conclusion without its quantifier. Steps 14–18 are concerned with putting the quantifier on by UG, and completing the existential specification used to eliminate the quantifier from the third premise.

It should be emphasized that UG and I16 and I17 can only be used with respect to an entire line—never just part of a line of a derivation. If for example, we had a derivation in which the sentence $[Fa \supset (\exists y)Gy]$ appeared, we could not obtain $[(x)Fx \supset (\exists y)Gy]$ by UG, nor could we obtain $[(\exists x)Fx \supset (\exists y)Gy]$ by I17. Similarly, if $[(x)Fx \supset (\exists y)Gy]$ appeared on a line of a derivation, we could not use I16 to obtain $[Fa \supset (\exists y)Gy]$. These restrictions on I16 and I17 are restrictions on Rule I in general. Remember that we can never use Rule I on just part of a line. And a similar restriction applies to Rule UG.

Clearly it will generally be much easier to show that a sentence is valid by constructing a derivation of it than by giving some sort of general argument involving truth under all interpretations. But, as in the case of PC derivations, the construction of derivations is not a purely mechanical matter. It will often require great ingenuity to find the appropriate derivation. In the light of this, one might desire some purely mechanical means, analogous to truth tables, for evaluating the validity or invalidity of sentences. Unfortunately, it can be proven that there is no such mechanical technique, so we will have to be content with using our ingenuity and constructing derivations.

III.14 EXERCISES

A. In each of the following purported derivations, check the lines for errors, and indicate the numbers of the lines on which errors occur. An error is said to occur in a line if *that* line cannot be introduced in accordance with the rules of inference. This means that in checking a line for errors, you only look at that line to see whether it has the necessary relation to the previous lines, and treat the previous lines as if they were all correct. If a line is incorrect, try to find a way of correcting it (perhaps by adding additional lines).

1.	(1)	1.	$(x)(Fx \vee Gx)$	P
	(2)	2.	$(\exists x)(y)(Fx \supset Gy)$	P
	(3)	3.	$(\exists x) \sim Gx$	P
	(2)	4.	$(y)(Fa \supset Gy)$	I, 2 (I17)
	(2)	5.	$(Fa \supset Gb)$	I, 4 (I16)
	(3)	6.	$\sim Gb$	P
	(2, 3)	7.	$\sim Fa$	I, 5, 6 (I10)
	(1)	8.	$(Fa \vee Ga)$	I, 1 (I16)
	(1, 2, 3)	9.	Ga	I, 7, 8 (I11)
	(1, 2, 3)	10.	$(x)Gx$	UG, 9
	(1, 2)	11.	$[(\exists x) \sim Gx \supset (x)Gx]$	C, 3, 10
	(1, 2)	12.	$[\sim (x)Gx \supset (x)Gx]$	R, 11 (E22)
	(1, 2)	13.	$[\sim \sim (x)Gx \vee (x)Gx]$	R, 12 (E4)
	(1, 2)	14.	$[(x)Gx \vee (x)Gx]$	R, 13 (E1)
	(1, 2)	15.	$(x)Gx$	R, 14 (E18)

2.	(1)	1.	$(\exists x)[Gx \supset (y)Hxy]$	P
	(2)	2.	$[Ga \supset (y)Hay]$	P
	(3)	3.	Ga	P
	(2, 3)	4.	$(y)Hay$	I, 2, 3 (I9)
	(2, 3)	5.	Hab	I, 4 (I16)
	(2)	6.	$(Ga \supset Hab)$	C, 3, 5
	(2)	7.	$(\exists y)(Gy \supset Hyb)$	I, 6 (I17)
	(2)	8.	$(x)(\exists y)(Gy \supset Hyx)$	UG, 7
		9.	$\{[Ga \supset (y)Hay] \supset (x)(\exists y)(Gy \supset Hyx)\}$	C, 2, 8
		10.	$\{(\exists x)[Gx \supset (y)Hxy] \supset (x)(\exists y)(Gy \supset Hyx)\}$	I, 9 (I17)
	(1)	11.	$(x)(\exists y)(Gy \supset Hyx)$	I, 1, 10 (I9)

B. For each of the following, construct a derivation of the sentence below the line from the sentences above the line. Keep the strategies in mind.

1. $$\begin{array}{l} (x)(Px \supset \sim Px) \\ \hline \sim Pa \end{array}$$

2. $$\begin{array}{l} (x)[Fx \supset (Hx \mathbin{\&} \sim Gx)] \\ \hline (Ga \supset \sim Fa) \end{array}$$

3. $$\begin{array}{l} (x)(Hx \supset Gx) \\ \hline [(Ha \supset Ga) \mathbin{\&} \sim(\sim Gb \mathbin{\&} Hb)] \end{array}$$

4. $$\begin{array}{l} (x)[Px \equiv (Hx \mathbin{\&} \sim Px)] \\ \hline (x) \sim Hx \end{array}$$

5. $$\begin{array}{l} (x)Fx \\ (x)(Fx \supset Gx) \\ \hline (x)Gx \end{array}$$

6. $$\begin{array}{l} Fa \\ \hline (\exists x)(Fx \vee Gx) \end{array}$$

7. $$\begin{array}{l} (\exists x)Fx \\ \hline (\exists x)(Fx \vee Gx) \end{array}$$

8. $$\begin{array}{l} (\exists x)(Fx \vee Gx) \\ \sim(\exists x)Fx \\ \hline (\exists x)Gx \end{array}$$

9. $$\begin{array}{l} [(\exists x)Fx \supset (y)Gy] \\ \hline (x)(y)(Fx \supset Gy) \end{array}$$

10. $$\begin{array}{l} (x)(Fx \supset Gx) \\ \sim(x)Gx \\ \hline (\exists y) \sim Fy \end{array}$$

11. $$\begin{array}{l} (x)[Fx \supset (Gx \supset Hx)] \\ \hline [(x)(Fx \supset Gx) \supset (x)(Fx \supset Hx)] \end{array}$$

12. $$\begin{array}{l} (x)(\exists y)Rxy \\ (x)(y)(Rxy \supset Ryx) \\ (x)(y)(z)[(Rxy \mathbin{\&} Ryz) \supset Rxz] \\ \hline (x)Rxx \end{array}$$

C. Show that each of the following sentences is valid by constructing a derivation of it; that is, a derivation whose last line has no premise numbers:

1. $[(x)Fx \supset (\exists x)Fx]$
2. $(\exists x)(Fx \supset (y)Fy)$ (*hint*: use reductio ad absurdum)
3. $[(x)(y)(Rxy \supset \sim Ryx) \supset \sim(\exists x)Rxx]$
4. $\sim(\exists x)(y)(Rxy \equiv \sim Ryy)$
5. $\sim(x)[(Fx \vee \sim Fx) \supset \sim(Fx \vee \sim Fx)]$
6. $\{(\exists x)(Fx \vee Gx) \equiv [(\exists x)Fx \vee (\exists x)Gx]\}$
7. $\{(x)(Fx \& Gx) \equiv [(x)Fx \& (x)Gx]\}$
8. $\{(x)(Fx \supset Gx) \supset [(x)Fx \supset (x)Gx]\}$
9. $\{(P \supset (x)Fx) \equiv (x)(P \supset Fx)\}$
10. $[(P \supset (\exists x)Fx) \equiv (\exists x)(P \supset Fx)]$
11. $[((x)Fx \supset P) \equiv (\exists x)(Fx \supset P)]$

D. Symbolize each of the following arguments in the predicate calculus, and construct a derivation of its conclusion from its premises.

1. Every teacher has some good students. Some students are also teachers. Thus some students have some good students.
2. Caesar was warned to beware the ides of March. If he had been warned by someone reliable, he would have been careful, but he was not careful. Therefore, he was warned by someone unreliable.
3. If someone were to steal the royal crown, he would not be able to sell it. If a person stole the royal crown and could not sell it, he would melt it down. If someone melts down the royal crown, he will discover that it is hollow, and know that the prince is a crook. Therefore, if someone steals the royal crown, then someone will know that the prince is a crook.
4. Given any positive integer, there is a greater positive integer. Therefore, given any positive integer, there are two positive integers such that the second is greater than the first, and the first is greater than the given integer.
5. Some people in England are older than anybody in Scotland. But some people in Siberia are older than anybody in England. If one person is older than a second person, and the second is older than a third, then the first person is older than the third. Therefore, some people in Siberia are older than anybody in Scotland.

E.* 1. As in the propositional calculus, our lists of implications and equivalences contain more schemes than necessary. Show that using only I16 and the second part of E22, together with the schemes from the propositional calculus, and all of the rules of inference, we can establish I17 (that is, derive the consequent of the implication from the antecedent), and the other three parts of E22, and E23–E27 (that is, construct a derivation of each side of each equivalence from the other side, showing that each side implies the other).

2. For some purposes it is of interest to construct concepts of validity, implication, and equivalence in which we allow interpretations to have empty domains. We adopt the convention that in such an interpretation all universal generalizations are true, and all existential generalizations are false.
   a. Is it possible to have a sentence that is not valid, but that would become valid if we allowed empty domains?
   b. Give an example of a sentence that is valid, but would not be valid if we allowed empty domains.
   c. Which of the implications I1–I17 and equivalences E1–E27 would cease to hold if we allowed empty domains?

III.15 ASSESSING ARGUMENTS

A Assessing Argument Forms

Thus far we have developed a complex apparatus for assessing the soundness of argument forms. Now let us look more closely at how it is to be applied. *If the form of an argument can be symbolized in the predicate calculus, then we can determine whether that argument form is sound by seeing whether the premises imply the conclusion.* This involves two things. In order to establish that the premises of the argument *do* imply the conclusion, we must construct a derivation in the predicate calculus of the conclusion from the premises. On the other hand, in order to establish that the premises *do not* imply the conclusion, we must construct an interpretation of the corresponding conditional under which the premises are all true and the conclusion false, thus demonstrating the invalidity of the corresponding conditional. If we do not know whether a particular argument form is sound, we may have to try both of these until we find a feasible derivation or interpretation.

Consider an example of how this might go. Suppose we are investigating the argument:

1. All Communists desire the overthrow of the government.
2. No member of the John Birch Society is a Communist.

---

3. Therefore, no member of the John Birch Society desires the overthrow of the government.

The form of this argument can be expressed in the predicate calculus as follows:

1. $(x)(Cx \supset Ox)$
2. $\sim(\exists x)(Bx \ \& \ Cx)$

---

3. $\sim(\exists x)(Bx \ \& \ Ox)$

The question is whether this argument form is sound. Your logical intuitions might tell you immediately whether it is sound, in which case you would only have to resort to the tools of the predicate calculus to confirm your intuitions. But if you do not know, or are not sure, then you must either give a derivation of the conclusion from the premises, thus showing that it is sound, or you must give an interpretation of the corresponding conditional under which the premises are all true and the conclusion is false, thus showing that it is not sound. Let us begin by trying to construct a derivation of the conclusion from the premises:

(1)	1.	$(x)(Cx \supset Ox)$	P
(2)	2.	$\sim(\exists x)(Bx \ \& \ Cx)$	P
(2)	3.	$(x) \sim (Bx \ \& \ Cx)$	R, 2 (E22)
(2)	4.	$(x)(\sim Bx \vee \sim Cx)$	R, 3 (E3)
(1)	5.	$(Ca \supset Oa)$	I, 1 (I16)
(2)	6.	$(\sim Ba \vee \sim Ca)$	I, 4 (I16)
(2)	7.	$(\sim Ca \vee \sim Ba)$	R, 6 (E12)
(2)	8.	$(Ca \supset \sim Ba)$	R, 7 (E4)

But now our derivation seems to have run out of steam. $\sim(\exists x)(Bx \ \& \ Ox)$ is equivalent to $(x)(Bx \supset \sim Ox)$. The natural way to get this is by universal generalization from $(Ba \supset \sim Oa)$. But there seems to be no way to get $(Ba \supset \sim Oa)$ from $(Ca \supset Oa)$ and $(Ca \supset \sim Ba)$. This suggests that the argument form is not sound. Instead let us try to find an interpretation under which the premises are all true and the conclusion false. Let us begin with the simplest case—a one-element domain. Let the domain be $\{1\}$. Premise 2 is true if there is nothing in the domain that is a member of the extensions of both B and C. One way to assure this is to let the extension of C be the empty set. So let us assign $\varnothing$ as the extension of C, which makes the second premise true. The first premise is true if, and only if, $(Ca \supset Oa)$ is true under every a-variant of the interpretation. But there will only be one a-variant, the one assigning 1 to a. Ca is false under that interpretation, because the extension of C is $\varnothing$, and so does not contain the extension of a. Thus the conditional has a false antecedent, making the conditional true. So the first premise is also true if we assign $\varnothing$ as the extension of C. Notice that this is so regardless of what we assign as the extension of O. Now we want to make the conclusion false. The conclusion will be false if, and only if, $(\exists x)(Bx \,\&\, Ox)$ is true. This is true if, and only if, there is something in the domain that is in the extensions of both B and O. As 1 is the only element of the domain, this means that the extensions of B and O must both be $\{1\}$. Therefore, under the interpretation whose domain is $\{1\}$, and which assigns the extension $\varnothing$ to C and $\{1\}$ to both B and O, the premises are both true and the conclusion false, demonstrating that the argument form is not sound.

Clearly, it can be a very involved process to discover whether an argument form expressed in the predicate calculus is sound. You may first try to find a derivation, and fail. Then you may try to find an interpretation, and fail there too. Then you may go back and try again to find a derivation, and so on.

The situation is somewhat simpler if the argument form is expressed in the propositional calculus. If you cannot find a derivation of the conclusion from the premises, or an interpretation under which the premises are all true and the conclusion false, there is always one sure method you can use. That is to construct a truth table for the corresponding conditional. This may take a great deal of paper, but at least it will always give you an answer in the end. Unfortunately, there is no such sure method for the predicate calculus.

Having shown that the above argument *form* is unsound, can we then conclude that the argument itself is unsound? Not necessarily. An argument form is sound if, and only if, every argument having that form is sound. Thus if some arguments having that form are sound and others are unsound, the argument form will be unsound. Thus the above argument form may be unsound even though some arguments having that form are sound. In particular, the argument concerning Communists and members of the John Birch Society might be sound. Is there any way then that we can go on to show that this particular argument is unsound?

B Establishing the Irrationality of an Argument

There is no very good way to show that individual arguments are unsound, but there is something else that we can do. We are not generally interested in evaluating arguments in the abstract. For the most part, we are only interested in an argument that has actually been put forward in an attempt to establish some conclusion. Under these circumstances, we can always ask the person who proposed the argument why he thinks it is sound. If he cannot answer this question, then it is irrational for him to think that the argument does establish its conclusion, so we can simply ignore the argument. Unless there is some reason to think that the argument is sound, the argument gives no-one any reason to think that the conclusion is true. On the other hand, if the person who proposed the argument can give a reason for thinking that the argument is sound, then we can try to evaluate that reason. One reason that will frequently be given is that every argument having that form is sound; the argument *form* is itself sound. If this answer is given, and the form of the argument can be expressed in the predicate calculus, then we can apply our logical apparatus to see whether this claim is correct. If the person who proposed the argument to show that no member of the John Birch Society desires the overthrow of the government maintained that this argument was sound because its form was sound, we could show that his reason was incorrect by showing, as above, that the argument form is not sound. If he could not then give any other reason for thinking that the argument was sound, it would be irrational for him to continue to maintain that it estabiishes its conclusion. If he still continued to maintain that the argument was sound, without giving any reason, then we could simply dismiss his argument as being irrational.

However, many arguments are sound for reasons other than the soundness of their forms. So other kinds of reasons might also be given for thinking that an argument is sound. The most common kind of reason involves the addition of an analytic premise. Consider the following argument:

1. Frederick is the husband of Erika.
2. All of Frederick's wives are blondes.

---

3. Therefore, Erika is a blonde.

If we symbolize the form of this argument we get the following:

1. Hfe
2. $(x)(Wxf \supset Bx)$

---

3. Be

where H stands for "(1) is the husband of (2)," W stands for "(1) is a wife of (2)," B stands for "is blonde," f stands for "Frederick," and e stands for "Erika." It is easy to find an interpretation under which these premises are all true and the conclusion false, so this argument form is not sound. However, the argument itself is sound. The argument is sound because there is a certain relationship between the meaning of "husband" and the meaning of "wife." It is analytically true that if one person is the husband of a second person, then the second person is a wife of the first person. This can be symbolized as $(x)(y)(Hxy \supset Wyx)$. If we add this as a new premise to the argument, the form of the new argument becomes:

1. Hfe
2. $(x)(Wxf \supset Bx)$
3. $(x)(y)(Hxy \supset Wyx)$

---

4. Be

This new argument form is sound, as can be shown by the following derivation of the conclusion from the premises:

(1)	1.	Hfe	P
(2)	2.	$(x)(Wxf \supset Bx)$	P
(3)	3.	$(x)(y)(Hxy \supset Wyx)$	P
(2)	4.	$(Wef \supset Be)$	I, 2 (I16)
(3)	5.	$(y)(Hfy \supset Wyf)$	I, 3 (I16)
(3)	6.	$(Hfe \supset Wef)$	I, 5 (I16)
(2, 3)	7.	$(Hfe \supset Be)$	I, 4, 6 (I12)
(1, 2, 3)	8.	Be	I, 1, 7 (I9)

Because this argument form is sound, the initial argument was sound. The reason for this is that the additional premise that we added is analytically true, and so need not have been mentioned in the argument. Suppose the argument having a certain set of premises, $P_1, \ldots, P_n$, and the conclusion, Q, is sound. This is the same thing as for the corresponding conditional $[(P_1 \;\&\; \ldots \;\&\; P_n) \supset Q]$ to be analytically true. But now, suppose that one of the premises, say P_1, is analytically true. By E15, this corresponding conditional is equivalent to the conditional $\{P_1 \supset [(P_2 \;\&\; \ldots \;\&\; P_n \supset Q]\}$. Thus this latter conditional must also be analytically true. Clearly, if a conditional is analytically true, and its antecedent is analytically true, then its consequent is analytically true. Hence, if P_1 is analytically true, then $[(P_2 \;\&\; \ldots \;\&\; P_n) \supset Q]$ is analytically true. Thus the argument whose premises are $P_2, \ldots, P_n$ and whose conclusion is Q is also sound. Therefore, in general, we can always delete an analytically true premise from a sound argument without making the argument unsound.

This is a very common reason for thinking that an argument is sound. Frequently, an argument whose form is not sound is itself sound because we can add an analytically true premise to it which will then make its form sound, as in the following argument:

1. All unmarried men are neurotic.
2. Patrick is a bachelor.

---

3. Therefore, Patrick is neurotic.

If we symbolize the form of this argument, we get the following:

1. $(x)((Ux \& Mx) \supset Nx)$
2. Bp

---

3. Np

which is not sound. But if we add the analytic premise "All bachelors are unmarried men," symbolizing it as $(x)[Bx \supset (Ux \& Mx)]$, we get a new argument form that is sound.

Generally, when a person gives an argument, and we ask him why the argument is sound, he will give a reason that will involve the soundness of some argument form. He might reply that any argument of that form is sound or that if you add a certain analytic premise the resulting argument form will be sound. If that argument form can be symbolized in the predicate calculus, then we can evaluate his claim by seeing whether the premises of the symbolized argument do imply the conclusion. If they do not, then his reason for thinking that the argument is sound was a bad one, and so it would be irrational to try to support the conclusion using that argument. Therefore, although the predicate calculus does not provide us with a way to show that particular arguments are unsound, it does provide us with a way to show that a particular reason for thinking that a particular argument is sound is not a good reason. Consequently, the predicate calculus often gives us a way to decide whether it is rational to base a certain conclusion on a certain argument.

C *Limitations on Predicates and Relations*

However, the predicate calculus is no good at all for assessing the worth of an argument if the form of the argument cannot be expressed in the predicate calculus. We must be particularly wary about assuming too hastily that a particular locution in English stands for predicate or relation. Consider the following argument:

1. John thought he saw a ghost.

2. Therefore, there are ghosts.

Clearly, this argument is unsound. But if we cast it in the predicate calculus, letting G stand for "is a ghost," T stand for "(1) thought he saw (2)," and j stand for "John," we get the following:

1. $(\exists x)(Gx \,\&\, Tjx)$

2. $(\exists x)Gx$

This argument form is sound, as can be seen from the following derivation:

(1)	1.	$(\exists x)(Gx \,\&\, Tjx)$	P
(2)	2.	$(Ga \,\&\, Tja)$	P
(2)	3.	Ga	I, 2 (I1)
(2)	4.	$(\exists x)Gx$	I, 3 (I17)
	5.	$[(Ga \,\&\, Tja) \supset (\exists x)Gx]$	C, 2, 4
	6.	$(x)[(Gx \,\&\, Tjx) \supset (\exists x)Gx]$	UG, 5
	7.	$[(\exists x)(Gx \,\&\, Tjx) \supset (\exists x)Gx]$	R, 6 (E25)
(1)	8.	$(\exists x)Gx$	I, 1, 7 (I9)

What has gone wrong here? The argument form is sound, but the argument is not. What this indicates is that this is not really the form of the above argument. The difficulty is that, in spite of appearances, "(1) thought he saw (2)" is not a relation. Relations are *extensional*;[12] that is, if a and b are the same object, and a stands in some relation, R, to another object, c, then b must also stand in that relation to c. That is, $(x)(y)(z)[x = y \supset (Rxz \equiv Ryz)]$. Consider the relation "(1) slugged (2)." Suppose that Mickey Finnegan is the superintendent of schools. Then if Tom Jones slugged Mickey Finnegan, it follows that Tom Jones slugged the superintendent of schools. The extensionality of relations is built into our truth rules for the predicate calculus. We assign to each individual constant a denotation, and to each relation an extension. Then if two individual constants, say a and b, have the same denotation (like "Mickey Finnegan" and

[12] This is merely a stipulation concerning what we are going to call by the term "relation." It does not involve any deep philosophical insight into anything.

"the superintendent of schools"), then an atomic sentence involving one of them is true if, and only if, the atomic sentence that results from substituting the other constant for the first one is true. Thus, *Rac* is true if, and only if, the ordered pair whose first member is the denotation of *a* and whose second member is the denotation of *c* is in the extension of *R*. And as *a* and *b* have the same denotation, this is the same as the condition under which *Rbc* is true. Thus *Rac* and *Rbc* will be true under the same conditions.

Only relations can be symbolized by relation symbols, because other expressions might not have the same truth conditions. And all relations are extensional. However, "(1) thought he saw (2)" is not extensional. Let us suppose that the mayor is the man who robbed the bank, but the police chief does not know this. Then "The police chief thought he saw the mayor" may be true at the same time that "The police chief thought he saw the man who robbed the bank" is false. Thus "(1) thought he saw (2)" is not a relation, and so cannot be symbolized using a relation symbol. This is what went wrong in the argument about ghosts.

The terms that most frequently generate nonextensional locutions are the "epistemic" terms like "knows," "believes," "thinks," and so on. For example, "(1) is thinking about (2)," "(1) knows that (2) is the man," and "(1) believes that (2) is a criminal" are all nonextensional. However, most other locutions will be extensional, and so can be symbolized using relation symbols. The way to determine whether any particular locution does correspond to a relation is just to see whether the denotation of a name or referential expression (like "the girl with the red hair") is all that matters for the truth of the statement. If for example, we are given the statement "Roosevelt saved the economy," and asked whether "(1) saved the economy" is a relation (or predicate), we can test this as follows. This is a relation if, and only if, we cannot change the truth value of the statement by substituting for "Roosevelt" any other expression that also refers to Roosevelt; for example, "The thirty-second president of the United States saved the economy" is true if, and only if, "Roosevelt saved the economy" is true. Thus "(1) saved the economy" is extensional, and so it is a relation.

D Conclusions

If the form of an argument can be expressed in the predicate calculus, then we can use the methods of the predicate calculus to determine whether that argument form is sound. If it is sound, this shows that the argument itself is sound. If it is not sound, we cannot conclude that the original argument was unsound; but we can demand a reason for thinking that the argument was sound, and such a reason is generally that some other argument form, augmented with analytic premises, is sound. So we can still use the predicate calculus to test the rationality of basing a conclusion on a given argument.

Let us close this section by applying our methods to the analysis of a rather difficult argument that has occurred in another area of philosophy—ethics. *Ethical egoism* is that moral theory according to which a person ought always to seek his own pleasure and never do anything else. A very common argument for this theory is based upon a psychological premise called *psychological egoism*: Because of his psychological makeup, a person cannot do other than seek his own pleasure. It is frequently claimed that ethical egoism follows logically from psychological egoism. Now let us investigate whether the argument whose sole premise is the principle of psychological egoism, and whose conclusion is the principle of ethical egoism, is sound.

1. A person cannot do other than seek his own pleasure.

---

2. Therefore, a person ought always to seek his own pleasure and never do anything else.

The first step in evaluating this argument is to symbolize its form in the predicate calculus, and see whether the form itself is sound. The process of symbolizing this argument is somewhat involved. The premise can first be paraphrased as "Given anyone, x, x cannot do other than seek his own pleasure." Then it can be further paraphrased as "Given anyone, x, any act that x can perform is an act of seeking his own pleasure." Paraphrasing again, we get "(x)[if x is a person, then given anything y, if y is an act and x can perform y, then y is an act of x's seeking his own pleasure]." Letting P stand for "is a person," A for "is an act," C for "(1) can perform (2)," and S for "(1) is an act of (2)'s seeking his own pleasure," we can symbolize this as $(x)[Px \supset (y)((Ay \,\&\, Cxy) \supset Syx)]$.

The conclusion of this argument can be paraphrased as "Given anyone, x, given anything, y, if y is an act, then x ought to do y if, and only if, y is an act of x's seeking his own pleasure." This can then be symbolized as $(x)[Px \supset (y)[Ay \supset (Oxy \equiv Syx)]]$, where now we have let O stand for "(1) ought to do (2)." So the form of the argument becomes:

$$\begin{array}{ll} 1. & (x)(Px \supset (y)[(Ay \,\&\, Cxy) \supset Syx]) \\ \hline 2. & (x)(Px \supset (y)[Ay \supset (Oxy \equiv Syx)]) \end{array}$$

This argument form is not sound. The premise is true and the conclusion false under the interpretation whose domain is $\{1, 2\}$, and which assigns the extension $\{1\}$ to P, $\{2\}$ to A, $\{\langle 1, 2\rangle\}$ to both C and S, and $\varnothing$ to O.

It is however, often claimed that the initial argument is sound because we get a sound argument form when we add the analytic premise that "ought" implies "can"; that is, "Any act that a person ought to do is an act that he can do." This premise can be symbolized as $(x)(y)[(Px \,\&\, Ay) \supset (Oxy \supset Cxy)]$. The new argument form now becomes

$$\begin{array}{ll} 1. & (x)(Px \supset (y)[(Ay \,\&\, Cxy) \supset Syx]) \\ 2. & (x)(y)[(Px \,\&\, Ay) \supset (Oxy \supset Cxy)] \\ \hline 3. & (x)(Px \supset (y)[Ay \supset (Oxy \equiv Syx)]) \end{array}$$

However, this new argument form is still not sound. Using the same interpretation we used above, the premises are both true and the conclusion false. But we are on the right track, because part of the conclusion, namely, $(x)(Px \supset (y)[Ay \supset (Oxy \supset Syx)])$ is implied by the premises. This says that any act that a person ought to perform is an act of seeking his own pleasure. The derivation is as follows:

(1)	1.	$(x)(Px \supset (y)[(Ay \,\&\, Cxy) \supset Syx])$	P
(2)	2.	$(x)(y)[(Px \,\&\, Ay) \supset (Oxy \supset Cxy)]$	P
(1)	3.	$(Pa \supset (y)[(Ay \,\&\, Cay) \supset Sya])$	I, 1 (I16)
(2)	4.	$(y)[(Pa \,\&\, Ay) \supset (Oay \supset Cay)]$	I, 2 (I16)
(2)	5.	$[(Pa \,\&\, Ab) \supset (Oab \supset Cab)]$	I, 4 (I16)
(6)	6.	Pa	P
(1, 6)	7.	$(y)[(Ay \,\&\, Cay) \supset Sya]$	I, 3, 6 (I9)
(1, 6)	8.	$[(Ab \,\&\, Cab) \supset Sba]$	I, 7 (I16)
(9)	9.	Ab	P
(6, 9)	10.	$(Pa \,\&\, Ab)$	I, 6, 9 (I13)
(2, 6, 9)	11.	$(Oab \supset Cab)$	I, 10, 5 (I9)
(12)	12.	Oab	P
(2, 6, 9, 12)	13.	Cab	I, 11, 12 (I9)
(2, 6, 9, 12)	14.	$(Ab \,\&\, Cab)$	I, 9, 13 (I13)
(1, 2, 6, 9, 12)	15.	Sba	I, 8, 14 (I9)
(1, 2, 6, 9)	16.	$(Oab \supset Sba)$	C, 12, 15

(1, 2, 6)	17.	$[Ab \supset (Oab \supset Sba)]$	C, 9, 16
(1, 2, 6)	18.	$(y)[Ay \supset (Oay \supset Sya)]$	UG, 17
(1, 2)	19.	$(Pa \supset (y)[Ay \supset (Oay \supset Sya)])$	C, 6, 18
(1, 2)	20.	$(x)(Px \supset (y)[Ay \supset (Oxy \supset Syx)])$	UG, 19

Thus we can obtain from premises 1 and 2 the conclusion that any act that a person ought to perform is an act of seeking his own pleasure. But we must still find a way of obtaining the additional conclusion that any act of seeking one's own pleasure is an act that he ought to perform; that is, $(x)(Px \supset (y)[Ay \supset (Syx \supset Oxy)])$. In order to do this, philosophers have sometimes wanted to add the premise that no act is morally indifferent; that given any act and any person, either the person ought to do the act, or he ought to refrain from doing the act. We can paraphrase this as "Given any person, *x*, and given any act, *y*, either *x* ought to do *y*, or given any act *z*, if *z* is an act of refraining from doing *y*, then *x* ought to do *z*." This can then be symbolized as $(x)(y)[(Px \& Ay) \supset (Oxy \vee (z)[Rzy \supset Oxz])]$, where now *R* stands for "(1) is an act of refraining from doing (2)." We also need several analytic premises concerned with the notion of refraining from doing something. The first is that given any act, there is another act that is an act of refraining from doing the first, $(x)(Ax \supset (\exists y)Ryx)$. The second is that an act of refraining from doing something is itself an act, $(x)((\exists y)Rxy \supset Ax)$. Finally, we need the premise that, given any act, it cannot be the case that both that act and an act of refraining from doing that act are acts of seeking one's own pleasure, namely, "Given any person, *x*, and any act, *y*, given anything, *z*, if *z* is an act of refraining from doing *y*, then it is not the case that both *y* is an act of *x*'s seeking his own pleasure, and *z* is an act of *x*'s seeking his own pleasure." This is symbolized as

$$(x)(y)(z)[(Px \& Ay \& Rzy) \supset \sim(Syx \& Szx)]$$

The new argument now becomes:

1. $(x)[Px \supset (y)[(Ay \& Cxy \supset Syx])$
2. $(x)(y)[(Px \& Ay) \supset (Oxy \supset Cxy)]$
3. $(x)(y)[(Px \& Ay) \supset (Oxy \vee (z)[Rzy \supset Oxz])]$
4. $(x)(y)(z)[(Px \& Ay \& Rzy \supset \sim(Syx \& Szx)]$
5. $(x)(Ax \supset (\exists y)Ryx)$
6. $(x)((\exists y)Rxy \supset Ax)$

---

7. $(x)(Px \supset (y)[Ay \supset (Oxy \equiv Syx)])$

This argument form is now sound. A derivation of the conclusion from the premises is quite straightforward. We begin by taking the additional premises *Pa* and *Ab*, with the objective of deriving $(Oab \equiv Sba)$, conditionalizing and using UG to get our conclusion. We can derive $(Oab \supset Sba)$ just as we did in the above derivation, so all that remains is to derive $(Sba \supset Oab)$. Let us take *Sba* as a new premise. By 5, $(\exists y)Ryb$, so suppose *Rcb* (a premise we can use for existential specification). Then by 4, $(Sba \supset {\sim}Sca)$, so ${\sim}Sca$. By 6, *Ac*, so then by 1, ${\sim}Cac$. Then by 2, ${\sim}Oac$. Thus ${\sim}(Rcb \supset Oac)$, and so ${\sim}(z)[Rab \supset Oac]$. Then by 3, *Oab*. Using existential specification, we eliminate the premise *Sba*, so *Oab* follows from the premises *Pa*, *Ab*, and *Sba*, together with 1–6. Then conditionalizing, $(Sba \supset Oab)$ follows from *Pa*, *Ab*, and 1–6. So we get $(Oab \equiv Sba)$. Then by conditionalization and UG, we get the conclusion.

The new argument form is sound. If we agree that the new premises are analytic, then this shows that the derivation of ethical egoism from psychological egoism is sound. Of course, the analyticity of some of these premises might well be disputed, but that takes us outside the realm of formal logic.

This example illustrates how useful the predicate calculus can be in assessing complicated arguments. The derivation of ethical egoism from psychological egoism is sufficiently complex that a great many philosophers have been misled about its soundness. But using the precise tools of the predicate calculus, we can see just what premises are necessary in order to make the argument go through. Similar situations arise in any area of intellectual endeavor in which one encounters complicated arguments.

III.15 EXERCISES

A. Assess the soundness of each of the following argument forms by either deriving the conclusion from the premises or constructing an interpretation under which the premises are true and the conclusion false:

1. $(x)[Fx \supset (Gx \vee Hx)]$
   $(x)(Gx \supset Mx)$
   $(x)(Hx \supset Mx)$
   ${\sim}Ma$
   ―――――――――
   ${\sim}(x)Fx$

2. $(x)(\exists y)(Hxy \;\&\; {\sim}Hyx)$
   ―――――――――
   $(\exists y)(x)(Hxy \;\&\; {\sim}Hyx)$

3. $[(P \,\&\, Q) \supset R]$
$(\sim R \supset \sim S)$
$(S \equiv \sim Q)$
―――――――
$\sim P$

4. $(x)(Fx \supset Gxx)$
$(x)(y)[Gxy \supset (Fx \equiv Fy)]$
―――――――
$(\exists x) \sim Gxx$

B. Show that the following arguments are sound by adding analytic premises that will make the forms of the arguments sound. Show that the resulting argument forms are sound by deriving the conclusions from the premises.

1. Joe is the father of Herkimer. Herkimer is a boy. Therefore, Herkimer is a son of Joe.
2. Mathilda is Patrick's aunt. Priscilla is Patrick's sister. Therefore, any husband of Mathilda is an uncle of Priscilla.
3. Midnight is the fastest racehorse alive. But no animal can move as fast as a Ferrari. Therefore, Midnight is slower than a Ferrari.
4. Only the governor of Hertzogovia can live in the palace. No one who does anything illegal can be govenor of Hertzogovia. The Hertzogovian police never arrest anyone unless he is a murderer. Count Zarnov was arrested in Sofia and extradited to Hertzogovia. Therefore, Count Zarnov can never live in the palace.
5. *For Whom the Bells Toll* is one of the great works of English literature. No one but Hemingway could have written that book. Thus, *For Whom the Bells Toll* was written by Hemingway.
6. The Amazon is the longest river in the world. Fresh water sharks live in the Amazon. These sharks cannot live in salt water. Thus, the Amazon is a body of fresh water.

III.16 COMPLETENESS AND THE LÖWENHEIM THEOREM

The principal objective of this section is to establish that every valid sentence of the predicate calculus is a theorem of the predicate calculus, thus showing that derivations do provide a way of establishing for each valid sentence that it is valid. This is called the Completeness Theorem for the predicate calculus. In connection with this, we will also prove the Löwenheim theorem, which was used in Section IX.

The proof of the completeness theorem is quite complex. Our strategy will be to first prove the following theorem:

> If P is any sentence of the predicate calculus such that $\sim P$ is not a theorem of the predicate calculus, then there is an interpretation

of P under which P is true that has as its domain the set of nonnegative integers.

The completeness theorem follows easily from this. By contraposition, we see that if P is not true under any interpretation whose domain is the set of nonnegative integers, then $\sim P$ is a theorem of the predicate calculus. If Q is a valid sentence, then $\sim Q$ is not true under any interpretation at all, much less under one whose domain is the set of nonnegative integers, so $\sim\sim Q$ is a theorem. But then Q is also a theorem. So if Q is valid, it is a theorem.

In order to prove the above theorem, we will construct a sequence of conjunctions having the properties that (1) the given sentence P is the first conjunct of each conjunction, (2) each conjunction is formed by adding one more conjunct to the previous conjunction, and (3) for each conjunction, its negation is not a theorem. We will let A be the set of all sentences that appear as conjuncts of these conjunctions. It will be shown that the sentences in A satisfy all of the truth rules for the predicate calculus, that is, a conjunction is in A if and only if both of its conjuncts are in A, a disjunction is in A if and only if at least one of its disjuncts is in A, an existential generalization $(\exists x)P$ is in A if and only if at least one instance of the form $Sb(\frac{a}{x})P$ is in A, and so on. It will then be shown that because A satisfies these rules, an interpretation can be constructed which makes all of the sentences in A true and has as its domain the set of nonnegative integers. As the given sentence P is in A, it will follow that P is true under an interpretation whose domain is the set of nonnegative integers, thus proving the theorem.

Suppose P is a sentence of the predicate calculus such that $\sim P$ is not a theorem. Consider all sentences of the predicate calculus that contain no relation symbols that do not occur in P. (So, for example, if P is the sentence $(\exists x)(Rx \;\&\; Sax)$, we consider all sentences that contain no relation symbols but R and S.) Let us arrange these sentences in an infinite list $P_1, P_2, P_3, \ldots, P_i, \ldots$, such that the first sentence of the list (that is P_1) is our sentence P, and such that each time an existential generalization, such as $(\exists x)Q$, appears in the list, it is followed immediately by a sentence of the form $Sb(\frac{a}{x})P$ where a is an individual constant that has not occurred in Q or in any of the previous sentences in the list. Clearly the sentences can be arranged in such a list. Next we define a sequence $Q_1, Q_2, Q_3, \ldots, Q_i, \ldots$, of conjunctions as follows:

Let Q_1 be P_1 (that is, P).

Let Q_2 be $(Q_1 \,\&\, P_2)$ if $\sim(Q_1 \,\&\, P_2)$ is not a theorem.
$(Q_1 \,\&\, \sim P_2)$ if $\sim(Q_1 \,\&\, P_2)$ is a theorem.

Let Q_3 be $(Q_2 \,\&\, P_3)$ if $\sim(Q_2 \,\&\, P_3)$ is not a theorem.
$(Q_2 \,\&\, \sim P_3)$ if $\sim(Q_2 \,\&\, P_3)$ is a theorem.

And so on. In general, for any number, n,

Let Q_n be $(Q_{n-1} \,\&\, P_n)$ if $\sim(Q_{n-1} \,\&\, P_n)$ is not a theorem.
$(Q_{n-1} \,\&\, \sim P_n)$ if $\sim(Q_{n-1} \,\&\, P_n)$ is a theorem.

Each conjunction in this sequence results from adding one more conjunct to the previous conjunction. This sequence of conjunctions will be used to construct an interpretation under which P is true. But first we must prove a number of things about the conjunctions:

Theorem 1: Given any number, n, if Q_n is defined as above, then $\sim Q_n$ is not a theorem of the predicate calculus.

Proof: First notice that Q_1 is just P, and our initial supposition was that $\sim P$ is not a theorem of the predicate calculus. So the theorem is true for $n = 1$. Next it will be proven that for any number, say k, if $\sim Q_k$ is not a theorem, then $\sim Q_{k+1}$ is not a theorem. Suppose that $\sim Q_k$ is not a theorem. Then it cannot be the case that both $\sim(Q_k \,\&\, P_{k+1})$ and $\sim(Q_k \,\&\, \sim P_{k+1})$ are theorems, because suppose they both were. Then both $(Q_k \supset P_{k+1})$ and $(Q_k \supset \sim P_{k+1})$ would be theorems, because the above are tautologically equivalent to them (and so by the completeness of the propositional calculus, they are derivable from one another). But from those two conditionals we can derive the conditional $[Q_k \supset (P_{k+1} \,\&\, \sim P_{k+1})]$, and from this we can derive $\sim Q_k$, which is contrary to the supposition that $\sim Q_k$ is not a theorem. Therefore, it cannot be the case that both $\sim(Q_k \,\&\, P_{k+1})$ and $\sim(Q_k \,\&\, \sim P_{k+1})$ are theorems. If $\sim(Q_k \,\&\, P_{k+1})$ is not a theorem, then Q_{k+1} is $(Q_k \,\&\, P_{k+1})$, and so $\sim Q_{k+1}$ is not a theorem. On the other hand, if $\sim(Q_k \,\&\, P_{k+1})$ is a theorem, then Q_{k+1} is $(Q_k \,\&\, \sim P_{k+1})$, and we have just seen that if $\sim(Q_k \,\&\, P_{k+1})$ is a theorem, then $\sim(Q_k \,\&\, \sim P_{k+1})$ is not (that is, $\sim Q_{k+1}$ is not). So it is true in general that, for any number, k, if $\sim Q_k$ is not a theorem, then $\sim Q_{k+1}$ is not a theorem. But we have seen that $\sim Q_1$ is not a theorem, so then $\sim Q_2$ is not a theorem. But then, because $\sim Q_2$ is not a theorem, $\sim Q_3$ is not a theorem. And so on. So for each number n, $\sim Q_n$ is not a theorem of the predicate calculus.

Theorem 2: If for some number, n, the sentence P_n is an existential generalization, say $(\exists x)Q$, and Q_n is $(Q_{n-1} \,\&\, P_n)$, then Q_{n+1} is $[(Q_{n-1} \,\&\, P_n) \,\&\, P_{n+1}]$.

Proof: Suppose P_n is $(\exists x)Q$, and Q_n is $(Q_{n-1} \,\&\, P)$, that is,

$$(Q_{n-1} \,\&\, (\exists x)Q).$$

Then, by the way we arranged the original list of sentences, P_{n+1} is some sentence of the form $Sb\binom{a}{x}Q$, where a is some individual constant not occurring in Q or in any of the earlier sentences in the list. Then a does not occur in $(Q_{n-1} \,\&\, (\exists x)Q)$.

Now, suppose that Q_{n+1} is not $[(Q_{n-1} \,\&\, P_n) \,\&\, P_{n+1}]$. By the way we defined Q_{n+1}, this means that $\sim[(Q_{n-1} \,\&\, P_n) \,\&\, P_{n+1}]$ is a theorem of the predicate calculus, that is, $\sim[(Q_{n-1} \,\&\, (\exists x)Q) \,\&\, Sb\binom{a}{x}Q]$ is a theorem. But we can construct the following derivation:

(1)	1.	$\sim[(Q_{n-1} \,\&\, (\exists x)Q) \,\&\, Sb\binom{a}{x}Q]$	P
(1)	2.	$[\sim(Q_{n-1} \,\&\, (\exists x)Q) \vee \sim Sb\binom{a}{x}Q]$	R, 1 (E3)
(1)	3.	$[(Q_{n-1} \,\&\, (\exists x)Q) \supset \sim Sb\binom{a}{x}Q]$	R, 2 (E4)
(1)	4.	$(x)[(Q_{n-1} \,\&\, (\exists x)Q) \supset \sim Q]$	(where x does not occur free in Q_{n-1} or $(\exists x)Q)$ UG, 3
(1)	5.	$[(Q_{n-1} \,\&\, (\exists x)Q) \supset (x) \sim Q]$	R, 4 (E24)
(1)	6.	$[(Q_{n-1} \,\&\, (\exists x)Q) \supset \sim(\exists x)Q]$	R, 5 (E22)
(1)	7.	$[Q_{n-1} \supset ((\exists x)Q \supset \sim(\exists x)Q)]$	R, 6 (E15)
(1)	8.	$[Q_{n-1} \supset (\sim(\exists x)Q \vee \sim(\exists x)Q)]$	R, 7 (E4)
(1)	9.	$[Q_{n-1} \supset \sim(\exists x)Q]$	R, 8 (E18)
(1)	10.	$[\sim Q_{n-1} \vee \sim(\exists x)Q]$	R, 9 (E4)
(1)	11.	$\sim[Q_{n-1} \,\&\, (\exists x)Q]$	R, 10 (E3)

Thus $\sim(Q_{n-1} \,\&\, P_n)$ is derivable from $\sim[(Q_{n-1} \,\&\, P_n) \,\&\, P_{n+1}]$. Therefore, if the latter is a theorem, so is the former. Thus (by contraposition), if $\sim(Q_{n-1} \,\&\, P_n)$ is not a theorem, then

$$\sim[(Q_{n-1} \,\&\, P_n) \,\&\, P_{n+1}]$$

is not a theorem. Therefore, if Q_n is $(Q_{n-1} \,\&\, P_n)$, then Q_{n+1} is

$$[(Q_{n-1} \,\&\, P_n) \,\&\, P_{n+1}]$$

This proves Theorem 2.

Let A be the set of all sentences in our initial list (the P_i's) that are conjuncts of the conjunctions Q_i. For example, P is in A. And if Q_2 is $(P \,\&\, P_2)$, then P_2 is in A, or if Q_2 is $(P \,\&\, \sim P_2)$, then $\sim P_2$ is in A. The set A will be used to define an interpretation under which the sentence P is true. But first we must prove the following theorems about A:

Theorem 3: For any number, n, either P_n or $\sim P_n$ is in A.

Proof: Q_n is either $(Q_{n-1} \& P_n)$ or $(Q_{n-1} \& \sim P_n)$.

Theorem 4: For any number, n, $\sim P_n$ is in A if, and only if, P_n is not in A.

Proof: By Theorem 3, either P_n or $\sim P_n$ is in A, so all that remains is to show that they cannot both be in A. Notice that if two sentences are both in A, then there is some number, k, such that both sentences are conjuncts of Q_k. This is because first one sentence will be introduced as a conjunct of some Q_i (for some number, i), and then the other sentence will be introduced as a conjunct for a later Q_j, but that Q_j will still have the first sentence as a conjunct. Thus, if both P_n and $\sim P_n$ are in A, then there is some number, k, such that both P_n and $\sim P_n$ are conjuncts of Q_k. But then $\sim Q_k$ will be a theorem of the predicate calculus. We saw in Theorem 1 that $\sim Q_k$ is not a theorem of the predicate calculus, so therefore our supposition that both P_n and $\sim P_n$ are in A must be false. So $\sim P_n$ is in A if, and only if, P_n is not in A.

Theorem 5: For any numbers, m and n, $(P_m \& P_n)$ is in A if, and only if, both P_m and P_n are in A.

Proof: Suppose $(P_m \& P_n)$ is in A. By Theorem 3, either P_m or $\sim P_m$ is in A. If $\sim P_m$ were in A, then there would have to be some number, k, such that both $(P_m \& P_n)$ and $\sim P_m$ were conjuncts of Q_k. But then $\sim Q_k$ would be a theorem of the predicate calculus, which, by Theorem 1, is not the case. Thus $\sim P_m$ cannot be in A, and so P_m must be in A. Analogously, P_n is in A. Therefore, if $(P_m \& P_n)$ is in A, then both P_m and P_n are in A.

Now suppose that both P_m and P_n are in A. By Theorem 1, either $(P_m \& P_n)$ or $\sim(P_m \& P_n)$ is in A (because this conjunction also appears in the initial list, and so is one of the P_i's). If $\sim(P_m \& P_n)$ is in A, then there must be some number, k, such that $\sim(P_m \& P_n)$, P_m, and P_n, are all conjuncts of Q_k. But then $\sim Q_k$ would be a theorem, which is impossible. So $\sim(P_m \& P_n)$ cannot be in A, and so $(P_m \& P_n)$ must be in A. Therefore, if P_m and P_n are both in A, then $(P_m \& P_n)$ is in A.

In an analogous manner we prove the following:

Theorem 6: For any numbers, m and n, $(P_m \vee P_n)$ is in A if, and only if, either P_m or P_n (or both) are in A.

Theorem 7: For any numbers, m and n, $(P_m \supset P_n)$ is in A if, and only if, either P_m is not in A or P_n is in A.

Theorem 8: For any numbers, m and n, $(P_m \equiv P_n)$ is in A if, and only if, either both P_m and P_n are in A, or neither P_m nor P_n is in A.

Now we need the following two theorems:

Theorem 9: For any number, n, if P_n is an existential generalization of the form $(\exists x)Q$, then P_n is in A if, and only if, there is some individual constant, say a, such that $Sb(\frac{a}{x})Q$ is in A.

Proof: Suppose there is some individual constant, say a, such that $Sb(\frac{a}{x})Q$ is in A. By Theorem 3, either $(\exists x)Q$ or $\sim(\exists x)Q$ is in A. If $\sim(\exists x)Q$ were in A, then there would be some number, k, such that both $Sb(\frac{a}{x})Q$ and $\sim(\exists x)Q$ were conjuncts of Q_k. But then $\sim Q_k$ would be a theorem (because of I17). So $\sim(\exists x)Q$ cannot be in A, and hence $(\exists x)Q$ must be. Therefore, if there is some individual constant, a, such that $Sb(\frac{a}{x})Q$ is in A, then $(\exists x)Q$ is in A.

Conversely, suppose $(\exists x)Q$ (which is P_n) is in A. Then Q_n must have been $(Q_{n-1}\ \&\ P_n)$, because that is the only way to get P_n into A if it is not a negation. But then, by Theorem 2, Q_{n+1} is $[(Q_{n-1}\ \&\ P_n)\ \&\ P_{n+1}]$, and so P_{n+1} is also in A. But P_{n+1} is some sentence of the form $Sb(\frac{a}{x})Q$. Therefore, if $(\exists x)Q$ is in A, then there is some individual constant, a, such that $Sb(\frac{a}{x})Q$ is in A.

Theorem 10: For any number, n, if P_n is a universal generalization of the form $(x)Q$, then P_n is in A if, and only if, given any individual constant, say a, $Sb(\frac{a}{x})Q$ is in A.

Proof: Suppose that $(x)Q$ is in A. Then given any individual constant, a, either $Sb(\frac{a}{x})Q$ or $\sim Sb(\frac{a}{x})Q$ is in A. But if $\sim Sb(\frac{a}{x})Q$ were in A, then there would be some number, k, such that both $(x)Q$ and $\sim Sb(\frac{a}{x})Q$ were conjuncts of Q_k. But then, by I16, $\sim Q_k$ would be a theorem. So $\sim Sb(\frac{a}{x})Q$ cannot be in A, and hence $Sb(\frac{a}{x})Q$ must be in A. Therefore, if $(x)Q$ is in A, then every sentence of the form $Sb(\frac{a}{x})Q$ is in A.

Conversely, suppose that, given any individual constant, say a, $Sb(\frac{a}{x})Q$ is in A. Either $(x)Q$ or $\sim(x)Q$ is in A. Suppose $\sim(x)Q$ is in A. Then either $(\exists x) \sim Q$ or $\sim(\exists x) \sim Q$ must be in A. If $\sim(\exists x) \sim Q$ were in A, then there would be some number, k, such that both $\sim(x)Q$ and $\sim(\exists x) \sim Q$ would be conjuncts of Q_k. But by E22, $\sim(x)Q$ is equivalent to $(\exists x) \sim Q$, so $\sim Q_k$ would be a theorem. Therefore,

$\sim(\exists x) \sim Q$ cannot be in A, and so $(\exists x) \sim Q$ must be in A. But then, by Theorem 10, there would be some sentence of the form $Sb(\frac{a}{x}) \sim Q$, which is the same sentence as $\sim Sb(\frac{a}{x})Q$, in A. But, by our initial supposition, $Sb(\frac{a}{x})Q$ must also be in A. Thus there must be some number, k, such that both $Sb(\frac{a}{x})Q$ and $\sim Sb(\frac{a}{x})Q$ are conjuncts of Q_k. But then $\sim Q_k$ would be a theorem, which is impossible. Thus our supposition that $\sim(x)Q$ is in A must be false. Hence, $(x)Q$ must be A. So if every sentence of the form $Sb(\frac{a}{x})Q$ is in A, then $(x)Q$ must be in A.

Using the properties that the set A has now been shown to have, a series of interpretations will be defined—one for each sentence in our initial list (which is a list of all sentences that can be constructed using no relation symbols other than those occurring in P). These interpretations will be identical except that some will assign denotations to individual constants to which others do not assign denotations, and extensions to relations symbols to which others do not assign extensions (depending upon whether that constant or relation symbol occurs in the sentence corresponding to the interpretation). Given any sentence in our list, say P_i, an interpretation of it can be constructed as follows. Let the domain be the set of nonnegative integers. Then associate each individual constant with an integer in the following way: a with 1, b with 2, c with 3, . . ., s with 19, a_1 with 20, b_1 with 21, . . ., s_1 with 38, a_2 with 39, b_2 with 40, . . ., s_2 with 57, a_3 with 58, and so on. Then for each individual constant occurring in P_i, assign as its denotation the number corresponding to it in the above way. For each n-place relation symbol, R, let the extension of R consist of all ordered n-tuples $\langle \alpha_1, \ldots, \alpha_n \rangle$ of numbers such that, if $a_1, \ldots, a_n$ are the individual constants associated with $\alpha_1, \ldots, \alpha_n$, then $Ra_1 \ldots a_n$ is a sentence in A. For each sentential letter, P, assign "true" to P if P is in A, and "false" to P otherwise.

Notice that if one sentence, say P, is part of another sentence, say Q, then P will be true under its interpretation if, and only if, it is also true under the interpretation of Q, because all of the relations symbols and individual constants in P will receive the same interpretation under both interpretations.

The connection between these interpretations and the set A is established by the following theorem:

Theorem 11: If P is any sentence in our list, then P is true under its interpretation (as defined above) if, and only if, P is in A.

Proof: First note that our interpretations have been so constructed that atomic sentences, that is, sentences that are either sentential letters or are of the form $Ra_1 \ldots a_n$, are true under their interpretations if, and only if, they are in A.

In order to proceed further we need the notion of the *order* of a formula of the predicate calculus. The order of a formula is defined to be the number of steps it takes to construct the formula using the rules for constructing them given in Section III.6. For example, an atomic formula is of order 1. The negation of an atomic formula is of order 2. Similarly, the conjunction of two atomic formulas is of order 3 (one step to construct each conjunct, and one step to conjoin them). Or, consider the sentence $(x)(Fx \vee \sim Gx)$. This is of order 5. Step one gives us Fx. Step two gives us Gx. Step three gives us $\sim Gx$. Step four gives us $(Fx \vee \sim Gx)$. And step five gives us $(x)(Fx \vee \sim Gx)$.

Using the notion of the order of a formula, what will be argued is that if the above theorem is true of all sentences of order less than some number, say n, then it is also true of all sentences of order n. From this it will follow that it is true of all sentences, for the following reason. We have already seen that the theorem is true of all sentences of order 1—that is, all atomic sentences. Therefore, the theorem is true of all sentences of order less than 2. Then it will follow that it is true of all sentences of order 2. As it is true of all sentences of order 1 and order 2, it is true of all sentences of order less than 3. So then it follows that it is true of all sentences of order 3. And so on, for each number n. So it is true of all sentences.

It must be shown that if the theorem is true of all sentences of order less than some number, n, then it is also true of all sentences of order n. Suppose then that the theorem is true of all sentences of order less than n, that is, suppose that given any sentence of order less than n, that sentence is true under its interpretation if and only if it is in A. We want to show that it follows from this that the same thing is true of any sentence of order n. Suppose we have some sentence of order n. We have seven cases to consider, depending on whether the sentence is a negation, conjunction, disjunction, conditional, biconditional, universal generalization, or existential generalization:

CASE 1. Consider a sentence $\sim P$ of order n. Then P is a sentence of order $n - 1$. By our hypothesis, P is true under its interpretation if, and only if, P is in A. $\sim P$ is true under its interpretation if, and only if, P is not true. Thus, $\sim P$ is true under its interpretation if, and only if, P is not in A. By Theorem 4, P is not in A if, and only if, $\sim P$ is in A. Thus $\sim P$ is true under its interpretation if, and only if, it is in A.

Case 2. Consider a sentence $(P \,\&\, Q)$ of order n. Then P and Q are both sentences or order less than n. $(P \,\&\, Q)$ is true under its interpretation if, and only if, both P and Q are true under their interpretations. As P and Q are of order less than n, they are true under their interpretations if, and only if, they are in A. P and Q are both in A if, and only if, $(P \,\&\, Q)$ is in A, by Theorem 5. So $(P \,\&\, Q)$ is true under its interpretation if, and only if, it is in A.

Case 3. Consider a sentence $(P \vee Q)$. This is analogous to Case 3, using Theorem 6.

Case 4. Consider a sentence $(P \supset Q)$. This is analogous to Case 2, using Theorem 7.

Case 5. Consider a sentence $(P \equiv Q)$. This is analogous to Case 2, using Theorem 8.

Case 6. Consider a sentence $(\exists x)Q$ of order n. This is true under its interpretation if, and only if, $Sb(\frac{a}{x})Q$ is true under some a-variant of that interpretation, that is, if, and only if, there is something we can assign as the denotation of a which will make $Sb(\frac{a}{x})Q$ true. But everything in the domain is the denotation of some individual constant, so this is the same as saying that there is some individual constant, say b, such that $Sb(\frac{b}{x})Q$ is true under its interpretation. $Sb(\frac{b}{x})Q$ is a sentence of order $n - 1$, so by our supposition, it is true under its interpretation if, and only if, it is in A. Thus $(\exists x)Q$ is true under its interpretation if, and only if, there is some individual constant, say b, such that $Sb(\frac{b}{x})Q$ is in A. But by Theorem 9, this is true if, and only if, $(\exists x)Q$ is in A. Thus $(\exists x)Q$ is true under its interpretation if, and only if, it is in A.

Case 7. Consider a sentence $(x)Q$ of order n. This is true under its interpretation if, and only if, $Sb(\frac{a}{x})Q$ is true under every a-variant of that interpretation, that is, if, and only if, no matter what we assign as the denotation of a, $Sb(\frac{a}{x})Q$ will be true. But everything in the domain is the denotation of some individual constant, so this is the same as saying that given any individual constant, say b, $Sb(\frac{b}{x})Q$ will be true under its interpretation. But $Sb(\frac{b}{x})Q$ is a sentence of order $n - 1$, so it is true under its interpretation if, and only if, it is in A. Thus $(x)Q$ is true under its interpretation if, and only if, for every individual constant b, $Sb(\frac{b}{x})Q$ is in A. But by Theorem 10, this is true if, and only if, $(x)Q$ is in A. Thus $(x)Q$ is true under its interpretation if, and only if, it is in A.

So we have shown in general that, if the theorem is true of all sentences of order less than n, then it is true of all sentences of order n. And, as we have seen, it follows from this that it is true of all sentences. As a corollary of Theorem 11, we see that the original sentence with which we began our construction, P, is true under its interpretation, because it is in A. So we have proven:

Theorem 12: If P is any sentence of the predicate calculus such that $\sim P$ is not a theorem of the predicate calculus, then there is an interpretation of P under which P is true that has as its domain the set of nonnegative integers.

Now we can prove the completeness theorem:

Completeness Theorem for the Predicate Calculus: All valid sentences of the predicate calculus are theorems of the predicate calculus.

Proof: Consider an arbitrary valid sentence P. As P is valid, $\sim P$ is not true under any interpretation. If $\sim \sim P$ were not a theorem of the predicate calculus, then by Theorem 12, $\sim P$ would be true under some interpretation. Thus, if $\sim P$ is not true under any interpretation, $\sim \sim P$ must be a theorem of the predicate calculus. But if $\sim \sim P$ is a theorem, so is P. Therefore, if P is valid then P is a theorem of the predicate calculus.

The completeness theorem shows that derivations really do give us a way of establishing that any valid sentence is valid. Thus it is never necessary to resort to the rather obscure general arguments about truth under all interpretations such as those in Section III.9.

The Löwenheim Theorem was used in Section III.9 to establish the connection between validity and formal analyticity. That theorem is generally formulated as follows:

Löwenheim Theorem: A sentence is consistent if, and only if, it is true under some interpretation whose domain is the set of nonnegative integers.

This is not the formulation of the theorem that was used in Section III.9, but let us prove it first. It follows trivially from the completeness theorem. If a sentence is true under some interpretation whose domain is the set of integers, then it is consistent. So let us prove the converse. If P is a consistent sentence, then $\sim P$ is invalid. We saw earlier that all theorems of the predicate calculus are valid, so $\sim P$ cannot be a theorem. Then by Theorem 12, P is true under some interpretation whose domain is the set of integers.

In order to prove the version of the Löwenheim Theorem that was used in Section III.9, we must first prove the following theorem:

Theorem 13: If a sentence, P, is true under some interpretation, then given any domain which is the same size or larger than the domain of this interpretation, there is another interpretation having this new domain under which P is true.

Proof: Suppose the original domain is the set A, and the new domain, which is either the same size or larger than the original, is B. Then we associate each element of A with an element of B in such a way that different elements are associated with different elements. If $k \in A$, let k' be the element of B that we associate with k. If B is larger than A, there will be some elements of B not associated with any element of A. Call the set of these elements B_o. Next, pick some element of A, and call it k_o. Then we associate each element of B with an element of A in such a way that, if k is in B but not in B_o, then the element of A associated with k is the element with which k was originally associated (that is, k is associated with an element m if, and only if, k is m'), and if k is in B_o, then k is associated with k_o. Thus all of the elements of B except those in B_o are associated with distinct elements of A, and all of the elements of B_o are associated with k_o. If $k \in B$, let k^* be the element of A associated with k. (Thus if $k \in A$, then $(k')^*$ is k.)

Now let us define a new interpretation of P as follows. Call the original interpretation I, and the new one I'. I' will have the domain B. For each individual constant, say a, if I assigns the denotation k to a, then I' assigns the denotation k'. For each predicate letter, say F, if I assigns a set of objects, say C, as the extension of F, then I' assigns the set of all members, k, of B, such that $k^* \in C$. In other words, the new extension is the set of all members of B that are associated with elements of A that are in the original extension. Analogously, for each n-place relation symbol, say R, if I assigns a set of n-tuples of members of A as the extension of R, then I' assigns the extension consisting of all n-tuples $\langle k_1, \ldots, k_n \rangle$ of members of B such that $\langle k_1^*, \ldots, k_n^* \rangle$ is an n-tuple in the original extension. Each sentential letter is assigned the same truth value under I' as it had under I.

For an example of this construction, suppose we begin with an interpretation having the domain $\{1, 2\}$ which assigns the denotation 1 to a and the extension $\{\langle 1, 2 \rangle, \langle 2, 1 \rangle\}$ to R. Now let us construct an interpretation having the larger domain $\{5, 6, 7, 8\}$. Let $1'$ be 5 and let $2'$ be 6. Let 2 be k_o. Then 5^* is 1, and 6^*, 7^*, and 8^* are all 2.

Then the new denotation of a is $1'$, which is 5. The new extension of R is $\{\langle 5, 6\rangle, \langle 5, 7\rangle, \langle 5, 8\rangle, \langle 6, 5\rangle, \langle 7, 5\rangle, \langle 8, 5\rangle\}$.

It must now be shown that P is true under I' if, and only if, it is true under I. In order to do this, we need two new concepts. First, let us say that a sentence which contains some number, say n, of distinct individual constants, is an *n-formula*. So, for example, $[Fab \supset (\exists y)Hyc]$ is a 3-formula. Now we want to define the notion of an ordered n-tuple *satisfying* an n-formula under an interpretation. Given an n-formula, Q, and an interpretation which contains an interpretation of Q (that is, which assigns truth values, extensions, and denotations, to all of the symbols of Q), if $\langle a_1, \ldots, a_n\rangle$ is an n-tuple of elements of the domain of the interpretation, then $\langle a_1, \ldots, a_n\rangle$ *satisfies* Q under this interpretation if, and only if, Q is true under a new interpretation that assigns the same extensions to relation symbols and the same truth values to sentential letters as this one, but assigns a_1 as the denotation of the first individual constant in Q (listed alphabetically), a_2 to the second, and so forth.

Finally it will be argued that for any n-formula Q which contains no relation symbols or sentential letters not contained in P, an n-tuple $\langle a_1, \ldots, a_n\rangle$ of elements of B satisfies Q under I' if, and only if, $\langle a_1^*, \ldots, a_n^*\rangle$ satisfies Q under I; or if Q contains no individual constants, then Q is true under I' if, and only if, Q is true under I. Clearly this is true for atomic formulas because of the way we have defined the extensions of the relation symbols under I'. So now it will be argued that, for any number, say n, if this is true for all sentences of order less than n, then it is also true for all sentences of order n. From this it follows that it is true for all sentences of any order, and so, in particular it is true of P. So, given some number, n, suppose this is true for all sentences of order less than n, and let us show that it must then also be true for all sentences of order n. We have seven cases to consider:

CASE 1. Consider a negation, $\sim Q$, of order n. Then Q is a sentence of order $n - 1$. If Q contains no individual constants, then $\sim Q$ is true under I' if, and only if, Q is false under I'; and $\sim Q$ is true under I if, and only if, Q is false under I. By our supposition, Q is false under I' if, and only if, Q is false under I. So $\sim Q$ is true under I' if, and only if, $\sim Q$ is true under I.

On the other hand, suppose Q is an n-formula. $\langle a_1, \ldots, a_n\rangle$ satisfies $\sim Q$ under I' if, and only if, $\langle a_1, \ldots, a_n\rangle$ does not satisfy Q under I'; and $\langle a_1^*, \ldots, a_n^*\rangle$ satisfies $\sim Q$ under I if, and only if, it does not satisfy Q under I. But Q is of order $n - 1$, so by our supposition,

$\langle a_1, \ldots, a_n \rangle$ satisfies Q under I' if, and only if, $\langle a_1^*, \ldots, a_n^* \rangle$ satisfies Q under I. So $\langle a_1, \ldots, a_n \rangle$ satisfies $\sim Q$ under I' if, and only if, $\langle a_1^*, \ldots, a_n^* \rangle$ satisfies $\sim Q$ under I.

CASE 2. Consider a conjunction $(R \& Q)$ of order n. If $(R \& Q)$ contains no individual constants, then the argument to show that it is true under I if, and only if, it is true under I', is analogous to Case 1. Suppose on the other hand that $(R \& Q)$ is an n-formula, and both R and Q contain individual constants. Consider an n-tuple $\langle a_1, \ldots, a_n \rangle$ of elements of B (the domain of I'). Suppose the individual constants occurring in $(R \& Q)$, listed alphabetically, are $b_1, \ldots, b_n$. Then let $b_1^\# = a_1$, $b_2^\# = a_2$, and so on. Suppose $c_1, \ldots, c_k$ are the individual constants occurring in R, and $d_1, \ldots, d_m$ are the individual constants occurring in Q. Then $\langle a_1, \ldots, a_n \rangle$ satisfies $(R \& Q)$ under I' if, and only if, $\langle c_1^\#, \ldots, c_k^\# \rangle$ satisfies R under I' and $\langle d_1^\#, \ldots, d_m^\# \rangle$ satisfies Q under I'; and analogously, $\langle a_1^*, \ldots, a_n^* \rangle$ satisfies $(R \& Q)$ under I if, and only if, $\langle (c_1^\#)^*, \ldots, (c_k^\#)^* \rangle$ satisfies R under I and $\langle (d_1^\#)^*, \ldots, (d_m^\#)^* \rangle$ satisfies Q under I. Both R and Q are of order less than n, so $\langle c_1^\#, \ldots, c_k^\# \rangle$ satisfies R under I' if, and only if, $\langle (c_1^\#)^*, \ldots, (c_k^\#)^* \rangle$ satisfies R under I; and $\langle d_1^\#, \ldots, d_m^\# \rangle$ satisfies Q under I' if, and only if, $\langle (d_1^\#)^*, \ldots, (d_m^\#)^* \rangle$ satisfies Q under I. Thus $\langle a_1, \ldots, a_n \rangle$ satisfies $(R \& Q)$ under I' if, and only if, $\langle a_1^*, \ldots, a_n^* \rangle$ satisfies $(R \& Q)$ under I.

Suppose instead that $(R \& Q)$ is an n-formula, but Q contains no individual constants. Then R must be an n-formula. Then $\langle a_1, \ldots, a_n \rangle$ satisfies $(R \& Q)$ under I' if, and only if, $\langle a_1, \ldots, a_n \rangle$ satisfies R and Q is true under I'; and $\langle a_1^*, \ldots, a_n^* \rangle$ satisfies $(R \& Q)$ under I if, and only if, $\langle a_1^*, \ldots, a_n^* \rangle$ satisfies R and Q is true under I. But R and Q are of order less than n, so $\langle a_1, \ldots, a_n \rangle$ satisfies R under I' if, and only if, $\langle a_1^*, \ldots, a_n^* \rangle$ satisfies R under I; and Q is true under I' if, and only if, Q is true under I. Therefore, $\langle a_1, \ldots, a_n \rangle$ satisfies $(R \& Q)$ under I' if, and only if, $\langle a_1^*, \ldots, a_n^* \rangle$ satisfies $(R \& Q)$ under I.

The argument is analogous if $(R \& Q)$ is an n-formula and R contains no individual constants.

CASES 3–5. $(R \vee Q)$, $(R \supset Q)$, $(R \equiv Q)$, analogous to Case 2.

CASE 6. Consider an existential generalization $(\exists x)Q$ of order n. Suppose $(\exists x)Q$ contains no individual constants. Then if a is an individual constant not assigned a denotation by I (and so not by I' either), $(\exists x)Q$ is true under I if, and only if, $Sb(\frac{a}{x})Q$ is true under some a-variant of I, that is, if and only if, some element α of A satisfies

$Sb(\frac{a}{x})Q$ under I. Analogously, $(\exists x)Q$ is true under I' if, and only if, some element β of B satisfies $Sb(\frac{a}{x})Q$ under I'. But $Sb(\frac{a}{x})Q$ is of order $n-1$, so if β satisfies $Sb(\frac{a}{x})Q$ under I' then β^* satisfies $Sb(\frac{a}{x})Q$ under I, and if α satisfies $Sb(\frac{a}{x})Q$ under I, then α' satisfies $Sb(\frac{a}{x})Q$ under I'. Thus some element of B satisfies $Sb(\frac{a}{x})Q$ under I' if, and only if, some element of A satisfies $Sb(\frac{a}{x})Q$ under I. Therefore, $(\exists x)Q$ is true under I' if, and only if, $(\exists x)Q$ is true under I.

If $(\exists x)Q$ is an n-formula, the argument is analogous.

CASE 7. $(x)Q$, analogous to Case 6.

Therefore, given any sentence, Q, which contains no relation symbols not occurring in P: if Q contains no individual constants, then Q is true under I if, and only if, Q is true under I'; and if Q is an n-formula, then $\langle a_1, \ldots, a_n\rangle$ satisfies Q under I' if, and only if, $\langle a_1^*, \ldots, a_n^*\rangle$ satisfies Q under I. Now, consider P. If P contains no individual constants, then P is true under I if, and only if, P is true under I'. Suppose, on the other hand, that P contains some number, say n, individual constants. Suppose a_1 is the denotation assigned to the first individual constant in P by I', a_2 is the denotation assigned to the second individual constant, and so on. Then P is true under I' if, and only if, $\langle a_1, \ldots, a_n\rangle$ satisfies P under I'; and P is true under I if, and only if, $\langle a_1^*, \ldots, a_n^*\rangle$ satisfies P under I. Thus P is true under I' if, and only if, P is true under I. This proves Theorem 13.

Now we can prove the formulation of the Löwenheim Theorem that we used in Section III.9:

Löwenheim Theorem (second formulation): If a sentence of the predicate calculus is true under every interpretation having some particular infinite domain, then the sentence is valid.[13]

Proof: By the first formulation of the Löwenheim Theorem, a sentence, $\sim P$, is consistent, if and only if, $\sim P$ is true under some interpretation whose domain is the set of nonnegative integers. Thus, P is true under every interpretation whose domain is the set of nonnegative integers if, and only if, $\sim P$ is inconsistent, that is, if, and only if, P is valid. Now, consider some particular infinite domain D. As D is infinite, it is larger than or the same size as the set of all nonnegative integers, so by Theorem 13, if $\sim P$ were true under some interpretation

[13] The second formulation of the Löwenheim Theorem is not equivalent to the first formulation. It is actually a bit stronger than the first formulation.

whose domain is the set of nonnegative integers, then $\sim P$ would be true under some interpretation whose domain is D. Thus, if P is true under every interpretation whose domain is D, then $\sim P$ cannot be true under any interpretation whose domain is the set of nonnegative integers, and so P must be true under every interpretation whose domain is the set of nonnegative integers. But then, P must be valid.

III.16 EXERCISE

Using the results of the exercises on page 136, prove that all derivations in the predicate calculus are sound. The proof is completely analogous to the corresponding proof in the propositional calculus.

IV OTHER TOPICS IN LOGIC

IV.1 INTRODUCTION

The propositional and predicate calculi form the basis for most of modern logic. Once the student has a familiarity with these fundamental theories, he can profitably discuss other topics in logic. There are logical theories that deal with statement and argument forms other than those studied in the propositional and predicate calculi. Also of considerable importance is the examination of some of the presuppositions of the above treatment of the propositional and predicate calculi. The latter takes us into what is properly called the philosophy of logic. This chapter will constitute a rather cursory introduction to some of these topics.

IV.2 USE AND MENTION

There are two things one may do with a linguistic expression—one may use it to talk about something else, or one may use another expression to say something about it. For example, we may use the sentence "The barn is red" to talk about the barn. Or we may wish to say something about this sentence itself—for example, that it contains four words. In the first case, we are *using* the sentence to make a statement about something, while in the second case we are *mentioning* the sentence in order to say something about it. This is the distinction between *use* and *mention*. Ordinarily of course we use sentences and other linguistic expressions to make statements about other things, but when we want to make statements about language itself, then we mention linguistic expressions.

In order to make a statement about something, we must employ either a name of the object talked about or a description of it. This remains true in making statements about language. If we want to say that the sentence "The barn is red" contains four words, we must have a way of naming this sentence. There is a standard convention for doing this. We form the name of a linguistic expression by enclosing that expression in quotation marks. We have already done this twice in this section with the sentence "The barn is red," and we just did it again in this sentence. In general, when constructing a sentence that mentions another linguistic expression, we must enclose the mentioned expression in quotation marks. This is called the *quotation name* of the expression. Consider the following sentences:

1. "The barn is red" contains four words.
2. "'The barn is red'" names the sentence 'The barn is red.'
3. Cats are animals.
4. Cats is the plural of cat.
5. "Cats" is the plural of "cat."
6. Washington is a state.
7. "Washington" is a state.
8. Washington is the name of a state.
9. "Washington" is the name of a state.

Of these, 1, 2, 3, 5, 6, and 9 are true; 7 and 8 are false; 4 is not even a grammatically correct sentence, because it has a plural subject and a singular verb. Let us look at 6–9; 6 says that Washington is a state, which is true; 7 says that the *word* "Washington" is a state, which is false; 8 says that the *state* Washington is the name of a state, which is false; 9 says that the word "Washington" is the name of a state, which is true. Notice that when a quotation name occurs within another set of quotation marks, as in 2, we alternate single and double quotes to make the expression easier to read.

When we use one language to talk about another language, the language we are talking about is called the *object language*, and the language we use to talk about it is the *metalanguage*. If we say "German is a complicated language," the object language is German, and the metalanguage is English. Notice that whether a language is an object language, a metalanguage, or neither, is always relative to a particular discussion. At one time we may use English to talk about German, and at another time we may use German to talk about English. Furthermore, the object language and the metalanguage need not be different languages; for example, we generally use English to talk about English.

The formulas and sentences of the propositional and predicate calculi are said to make up the "languages" of the propositional and predicate calculi. They are called *artificial languages* to contrast them with natural languages like English or French. However, it should be emphasized that they are not really languages at all in the ordinary sense of that word. Nor are the sentences of the predicate calculus really sentences. In a real language, sentences must have more or less fixed meanings, but in the artificial languages of the propositional and predicate calculi the sentences do not have fixed meanings. We give them meaning by specifying an interpretation, and we can give them different meanings on different occasions.

Throughout this book we have talked about the artificial languages of the propositional and predicate calculi, so they have been object languages. Our metalanguage has been English. When we say something in English about an expression in one of the artificial languages, we are mentioning that expression and should properly enclose it in quotation marks. For example, when we wrote in the middle of page 13, "Then by Rule 2, $\sim Q$ is a formula," this was incorrect. We should have written, "Then by Rule 2, '$\sim Q$' is a formula," enclosing "$\sim Q$" in quotation marks. Throughout this book, when sentences of the metalanguage were written about specific formulas or sentences of the object language, we have tended to omit the quotation marks around the latter expressions. This is not correct according to the convention mentioned above. However, it is a rather common practice, and it seems like a good idea to adopt the convention of letting a formula of the object language be an abbreviation for its own quotation name. In other words, we will adopt the convention of omitting the parentheses around mentioned expressions of the object language provided no confusion is likely to result. Following this convention, we can write "$\sim Q$ is a formula," but we must be aware that this is only an abbreviation for "'$\sim Q$' is a formula."

We frequently have occasion in the metalanguage to make general statements about expressions of the object language. For example, we may want to say that given any sentence of the object language, the result of appending "$\sim$" to the front of it is another sentence. To make these statements precise, we need variables in the metalanguage that range over expressions in the object language. These are called *metalinguistic variables*. Throughout Chapters II and III we used metalinguistic variables without explicitly mentioning them. "P" and "Q" were used not only as atomic sentences, but also as metalinguistic variables ranging over formulas of the propositional and predicate calculi. For example, on page 107, in formulating our rules for constructing formulas of the predicate calculus, we wrote

2. If P is a formula, then $\sim P$ is a formula.

"P" was serving as a metalinguistic variable. Similarly, on page 109 we used "α" and "β" as metalinguistic variables ranging over the class of individual symbols. This illicit use of metalinguistic variables was hard to avoid at the time, because the student was not yet familiar with the concept of a variable, but now we can be quite precise about this. Henceforth, we will use the Greek letters "α" (alpha), "β" (beta), and "γ" (gamma), with or without numerical subscripts, as metalinguistic variables ranging over individual symbols, and we will use the Greek letters "φ" (phi), "θ" (theta), and "ψ" (psi), with or without numerical subscripts, as metalinguistic variables ranging over formulas.

Let us see how, using metalinguistic variables, we can state more precisely the rules given on page 107 for constructing formulas of the predicate calculus:

1. An atomic formula is a formula.
2. If φ is a formula, then the result of appending "$\sim$" to φ is a formula.
3. If φ and θ are formulas, then the result of writing "(," followed by φ, followed by either "&," "$\vee$," "$\supset$," or "$\equiv$," followed by θ, followed by ")" is a formula.
4. If φ is a formula and α is any variable, then the result of writing "(," followed by α, followed by ")," followed by φ, is a formula, and the result of writing "(," followed by "$\exists$," followed by α, followed by ")," followed by φ, is a formula.

The expression of these rules seems somewhat unwieldy. In order to simplify them one may be tempted to write, for example in place of 2:

2. If φ is a formula, then "$\sim\varphi$" is a formula.

or perhaps

2. If φ is a formula, then $\sim\varphi$ is a formula.

However, both of these are incorrect. The first alternative says that the result of appending "$\sim$" to the metalinguistic variable "φ" is a formula, but what we are trying to say is that the result of appending "$\sim$" to one of the objects over which the metalinguistic variable ranges is a formula. The second attempt is incorrect, because the expression "$\sim\varphi$" in the consequent of the conditional is not a name or variable, and so cannot be the subject of the clause.

There is no way to simplify these rules using ordinary quotation marks. The customary procedure is to introduce new symbols, "$\ulcorner$" and "$\urcorner$," called *quasi-quotes*. Then an expression of the form $\ulcorner\ldots\varphi\ldots\urcorner$ is defined to mean "the result of writing ... (whatever expression is in the place of the dots), followed by φ, followed by ...." For example, $\ulcorner(\varphi \supset \theta)\urcorner$ means "the result of writing '(,' followed by φ, followed by '$\supset$,' followed by θ, followed by ')'." Using quasi-quotes we can rewrite our rules as follows:

1. An atomic formula is a formula.
2. If φ is a formula, then $\ulcorner\sim\varphi\urcorner$ is a formula.
3. If φ and θ are formulas, then $\ulcorner(\varphi \;\&\; \theta)\urcorner$, $\ulcorner(\varphi \vee \theta)\urcorner$, $\ulcorner(\varphi \supset \theta)\urcorner$, and $\ulcorner(\varphi \equiv \theta)\urcorner$ are formulas.
4. If φ is a formula and α is any variable, then $\ulcorner(\alpha)\varphi\urcorner$ and $\ulcorner(\exists\alpha)\varphi\urcorner$ are formulas.

There are numerous places throughout the text where we should have used metalinguistic variables and quasi-quotes. However, just as we did with ordinary quotation marks, we will adopt the convention of omitting quasi-quotes provided no confusion is likely to result from doing so.

IV.2 EXERCISES

Which of the following are grammatically correct? Of those that are grammatically correct, which are true?

1. "Caesar" was an emperor of Rome.
2. Boston contains six letters.
3. Given any formula φ, $\ulcorner(P \vee \varphi)\urcorner$ is a disjunction, the first disjunct of which is an atomic formula.
4. Given any formula φ, $\ulcorner(\text{"}P\text{"} \vee \varphi)\urcorner$ is a disjunction, the first disjunct of which is an atomic formula.

5. "φ" is a metalinguistic variable.
6. $\ulcorner\varphi\urcorner$ is a metalinguistic variable.
7. φ is a metalinguistic variable.
8. "$\ulcorner\varphi\urcorner$" means the same thing as "φ."
9. "$\ulcorner\varphi\urcorner$" means the same thing as "'φ'."
10. "$\ulcorner P\urcorner$" means the same thing as "'P'."

IV.3 LOGICAL THEORIES

So far we have studied two logical theories—the propositional calculus and the predicate calculus. In the propositional calculus we considered those statement forms and argument forms that can be symbolized using the sentential connectives and sentential letters. In the predicate calculus we considered those statement forms and argument forms that can be symbolized using the sentential connectives, quantifiers, variables, individual constants, and relation symbols. Other logical theories can be generated by considering other statement and argument forms.

In constructing a logical theory we begin by choosing a set of symbols out of which our formulas will be constructed, and then specifying which strings of those symbols are sentences and formulas. In doing this we say we are constructing the *language of the theory*. We divide the symbols into two classes, the logical symbols and the nonlogical symbols. The difference between the two classes of symbols is that the logical symbols have the same meaning or function every time they occur, but the nonlogical symbols do not have fixed meanings. We must supply meanings for the nonlogical symbols each time we use them. In the propositional calculus for example, the sentential connectives and parentheses are logical symbols, while sentential letters are nonlogical symbols. Each time we use a sentential letter we must specify its meaning, but the sentential connectives and parentheses always have the same meaning in every formula. Analogously, in the predicate calculus the sentential connectives, parentheses, and variables are logical symbols, while the quantifiers, individual constants, and relation symbols are nonlogical symbols. The latter must be interpreted in each case by specifying a universe of discourse for the quantifiers, denotations for the individual constants, and meanings or extensions for the relation symbols.

In a logical theory, sentences express the forms of statements. We get a statement from a sentence by specifying the meanings of the nonlogical symbols occurring in the sentence. As in the propositional and predicate calculi, our interest is in sound argument forms and formally analytic statement forms. This leads us to seek a concept of an *interpretation* for our logical theory, and to give rules for determining what sentences are true under an interpretation, in such a way that we can show that a sentence of the theory is true under every interpretation if, and only if, every statement of that form is analytic. As in the predicate calculus, we say that a sentence is *valid* if, and only if, it is true under every interpretation. (In the propositional calculus, valid sentences were called tautologies.) And once again we define implication and equivalence in terms of the validity of the corresponding conditional and biconditional respectively.

It is useful to distinguish between two concepts, an interpretation of an individual sentence and an interpretation of the entire language. In our discussion of the predicate calculus we dealt only with interpretations of individual sentences, in which we specified extensions for the relation symbols occurring in the sentence and denotations for the individual constants occurring in the sentence, but we left other relation symbols and individual constants uninterpreted. On the other hand, in an interpretation of the language we interpret all of the nonlogical symbols—not only those in a particular sentence. The reason we dealt only with interpretations of particular sentences in our discussion of the predicate calculus was that we could display such interpretations completely. We could not do this for an interpretation of the language, because that involves interpretations of infinitely many symbols. On the other hand, there is sometimes an advantage in working with interpretations of the language rather than those of individual sentences because we can then talk about an interpretation making an infinite class of sentences true simultaneously. Had we had the concept of an interpretation of the language, this would have simplified somewhat the completeness proof for the predicate calculus. From the point of view of validity, it makes no difference which concept of an interpretation we use. A sentence is true under every interpretation of itself if, and only if, it is true under every interpretation of the language. This is because, for any given interpretation of the language, it is only that part of the interpretation that interprets the symbols occurring in that particular sentence (the part that constitutes an interpretation of that sentence) which has any effect on the truth of the sentence.

When we choose different concepts to symbolize, using symbols that we specify as logical symbols, we get different logical theories, because there are different classes of statement forms and argument forms that can be symbolized using those logical symbols. Let us consider briefly some examples of logical theories other than the propositional and predicate calculi.

a Restricted Propositional Calculi. In what we might call restricted propositional calculi, we take for our logical symbols parentheses and a restricted class of the sentential connectives. For example, we might consider just those formulas in which the only sentential connective is "$\supset$." Here we could define the notion of an interpretation in the same way as in the ordinary propositional calculus (as an assignment of truth values to sentential letters), and the truth rules would be the same except that the only connective mentioned would be "$\supset$." The valid sentences of this restricted propositional calculus would be all those tautologies in which the only sentential connective that occurs is "$\supset$." For example, "$[P \supset (Q \supset P)]$" would be valid.

Other restricted propositional calculi would result if we chose different sets of sentential connectives. For example, we might choose "$\sim$" and "&," or "$\sim$" and "$\supset$." It is interesting that were we to choose just "&," or just "$\vee$," or just "&" and "$\vee$," there would be no valid sentences.

b Modal Logic. In modal logic we add additional logical symbols to the propositional calculus. These are usually some or all of the symbols "$\Box$," "$\Diamond$," and "$\rightarrow$." These symbols are called *modal operators.* "$\Box$" and "$\Diamond$" are both "one-place" symbols, like "$\sim$," that are attached to the front of a sentence to obtain another sentence, while "$\rightarrow$" is a "two-place" symbol, like "&" and "$\vee$," that is put between two sentences to form another sentence. In other words, if φ is a sentence, then $\ulcorner\Box\varphi\urcorner$ and $\ulcorner\Diamond\varphi\urcorner$ are sentences, and if φ and θ are sentences, then $\ulcorner(\varphi \rightarrow \theta)\urcorner$ is a sentence.

The modal operators can be interpreted in a variety of ways, and depending upon how we interpret them we get different logical theories, all of which can be termed "modal logics." The most common is *alethic modal logic*, in which "$\Box$" is read "it is logically necessary that." For example, the statement "It is logically necessary that it is either raining or not raining" could be symbolized in alethic modal logic as "$\Box(P \vee \sim P)$." "$\Diamond$" is read "it is logically possible that," and "$\rightarrow$" symbolizes logical entailment, that is, "$(P \rightarrow Q)$" can be read as "The conclusion that Q can be inferred from the supposition that P."

In *epistemic logic*, "$\Box P$" is read "it is known that P," and "$\Diamond P$" is read "it is possible, for all that is known, that P." In *doxastic logic*, "$\Box P$" is read "It is believed that P," and "$\Diamond P$" is read "It is not disbelieved that P." In *deontic logic*, "$\Box P$" is read "It is morally obligatory that P," and "$\Diamond P$" is read "It is morally permissible that P."

Modern logicians have only recently become interested in modal logic, and it is still in a rather rudimentary state. There is little general agreement concerning how the concepts of an interpretation and validity should be defined for modal logic.

c The Monadic Predicate Calculus. If we eliminate from the predicate calculus all relation symbols of two places or more, the monadic predicate calculus is the theory that results. Thus in the monadic predicate calculus we are left with sentential letters and predicate letters as our only relation symbols. For example, "$(x)(Fx \supset P)$" is a sentence of the monadic predicate calculus, but "$(x)(\exists y)Rxy$" is not. It turns out that the structure of this logical theory is significantly simpler than that of the ordinary predicate calculus. We noted in passing that, unlike the propositional calculus (where we have truth tables), there is no mechanical test for the validity of sentences of the predicate calculus. However, in the monadic predicate calculus there is once again a mechanical test for validity.

d Quantifier Free Predicate Calculus. The quantifier free predicate calculus is the theory that results if we eliminate from the predicate calculus all quantifiers and variables. For example, "$[Rab \supset \sim(P \,\&\, Fb)]$" is a sentence of the quantifier free predicate calculus. A sentence of the quantifier free predicate calculus is valid if, and only if, it is a tautology.

e Second Order Predicate Calculus. In the ordinary predicate calculus our quantifiers range over individuals in the universe of discourse. In the second order predicate calculus we add quantifiers ranging over relations, and add appropriate variables. We will use "X," "Y," and "Z" with numerical superscripts and with or without numerical subscripts, as relation variables. The superscript will indicate the number of places in the relation. For example, if we write "$(X^0)(\exists Y^0)(X^0 \equiv \sim Y^0)$," this says that for any statement X^0, there is a statement Y^0 such that X^0 is true if, and only if, Y^0 is false. Similarly, "$(X^0)(\exists Y^2)(\exists x)(\exists y)(X^0 \equiv Y^2xy)$" says that given any statement X^0, there is a two-place relation Y^2 and there are individuals x and y, such that X^0 is true if, and only if, Y^2xy is true.

The concept of an interpretation is the same for the second order predicate calculus as for the ordinary predicate calculus, however we must add truth rules to tell us when sentences containing quantifiers over relations are true. Roughly, a sentence of the form $(X^n)\varphi$ is true under an interpretation if, and only if, $Sb(\frac{R^n}{X^n})\varphi$ would be true under that interpretation no matter what extension we assign to the n-ary relation symbol R^n.[1]

This can be made more precise in the same way that the corresponding rule for quantifiers over individuals was made precise in the ordinary predicate calculus. The analogous rule is used for existential quantifiers over relations.

One rather surprising fact about the second order predicate calculus is that there is no way to define the concept of a derivation which will result in all valid sentences being derivable. This means that, unlike the ordinary predicate calculus, there is no mechanical way to generate all valid sentences. Therefore the second order predicate calculus is much more difficult to study.

In the next two sections of this chapter we will study two logical theories somewhat more extensively—syllogistic and the predicate calculus with identity.

Given a logical theory, the problem arises of showing that particular sentences are valid or imply one another. This *can* be done by applying the theory's definition of validity directly and trying to show that particular sentences are valid. However, as in the case of the predicate calculus, this may be very difficult to do. Often we can formulate rules which can be followed in a mechanical manner and which are sufficient, ultimately, to generate any valid sentence and any implication. A sequence of sentences constructed according to these rules is called a *derivation.*

Derivations give us a relatively simple way of showing that sentences are valid, and for this purpose alone they are important. However, the construction of a system of derivations is often carried out with a second purpose in mind, too. An attempt is made to choose "natural" rules of inference so that the resulting derivations constitute natural patterns of inference. If this is done skillfully, derivations will provide

[1] Notice that we are using X and R here as metalinguistic variables.

us with a tool for analyzing our ordinary informal reasoning. The systems of derivations developed in Chapters II and III for the propositional and predicate calculi were intended to do this.

In order to accomplish what it is intended to do, a system of derivations must have two properties—it must be sound and complete. A system of derivations is said to be *sound* if, and only if, (1) all theorems are valid, and (2) whenever a sentence is derivable from a set of sentences it is implied by that set of sentences. A system of derivations is said to be *complete* if, and only if, (1) all valid sentences are theorems, and (2) whenever a sentence is implied by a set of sentences it is derivable from that set of sentences. Roughly speaking, a system of derivations is sound provided it does not give us anything we do not want, and it is complete if it does give us everything we do want. The derivations discussed for the propositional and predicate calculi were both sound and complete.

Two further concepts are of some importance in discussing logical theories. For some logical theories there is a mechanical test for determining the validity of a sentence; truth tables provided such a test for the propositional calculus. Given any formula of the propositional calculus, it is a purely mechanical matter of following precise rules, and requiring no ingenuity, to construct a truth table and determine whether the formula is a tautology. A theory for which there is such a mechanical test for validity is said to be *effectively decidable*. The mechanical test is called a *decision procedure*. The propositional calculus is effectively decidable but the predicate calculus is not. The theorem that the predicate calculus is not effectively decidable is called *Church's Theorem*. It was proven by Alonzo Church, and its proof is extremely complicated.

Related to the concept of effective decidability is the concept of effective generability. Although there is no mechanical test for validity in the predicate calculus, there is a set of rules that can be followed purely mechanically and which will ultimately generate any given valid sentence. They are the rules for constructing derivations. We could program a computer to construct derivations, and given any particular valid sentence, the computer would eventually construct a derivation of it. This gives us a mechanical way of showing that sentences *are* valid, but it does not give us a way of showing that sentences are *not* valid, and therefore does not constitute a decision procedure. This is because at any given time the fact that the computer has not yet constructed a derivation of a particular sentence does not show that it will not do so at some future time, and so does not show that the sentence is not valid. A logical theory for which there exists a mechanical way of generating all valid sentences is said to be *effectively generable*. The propositional and predicate calculi are both effectively generable, but it can be shown that the second order predicate calculus is not.

Two questions now arise in connection with arbitrary logical theories. First, how broadly can we choose our logical symbols? In various logical theories we have taken logical symbols to symbolize conjunction, disjunction, negation, and the other truth functional connectives; to symbolize quantifiers over classes of individuals, and quantifiers over classes of relations; and to symbolize logical necessity, logical possibility, entailment, and the concepts of being known, being believed, being morally obligatory, and being morally permissible. Is there anything that ties all of these concepts together, making them all "logical" concepts? If there is, then it would seem reasonable to require of a logical theory that the logical symbols only symbolize logical concepts. However, it is hard to see anything that all these concepts have in common. And even if there were some such characteristic that would make some concepts logical and others nonlogical, this would not mean that we could not construct theories in which we interpreted the logical symbols as symbolizing nonlogical concepts; it would just mean that we could not call such theories *logical* theories. Regardless of what we might call them, such theories could be quite interesting. For example, it has sometimes been maintained that, with the exception of alethic modal logic, modal logic is not logic, properly speaking. In other words, the concepts of knowledge, belief, and moral obligatoriness are not logical concepts. Even if this is true, it does not make these modal logics any less interesting. This suggests that, even if there is some distinction that can be made between logical and nonlogical concepts, it is not a very useful distinction for present purposes, and therefore we may as well call all of these theories logical theories, regardless of whether the logical symbols symbolize logical concepts.

The second question we must consider is whether there is such a thing as *the* form of a statement. We talked throughout Chapters II and III as if there were, but this now becomes questionable. It is natural to say that in order to obtain the form of a statement, we express it in terms of its simplest constituents and then replace the logical concepts by logical symbols and the nonlogical concepts by nonlogical symbols.

For example, in considering "Either it is raining or it is not raining," we might argue that it has three basic constituents—disjunction, negation, and the simple (not further analyzable) statement "It is raining." Symbolizing disjunction by "$\vee$," negation by "$\sim$," and "It is raining" by the nonlogical symbol "P" we obtain the statement form "$(P \vee \sim P)$." Any other statement that we can symbolize in this same way, letting "P" stand for something different, will have the same form.

However, there are two sources of difficulty with this procedure. The first concerns the question of whether there is a legitimate distinction between logical and nonlogical concepts. If there is not (and this author does not see how such a distinction can be made), then we cannot say what constituents are to be symbolized using logical symbols having a fixed meaning, and what constituents are to be symbolized using nonlogical symbols. If it is simply a matter of choice as to what we are going to treat as logical concepts at any given time, then different choices will lead to different symbolizations of the same statement, each of which has an equal right to be called the form of the statement.

The second difficulty concerns whether or not there are such things as the *simplest* constituents of a statement. It is plausible to suppose that each constituent can be further analyzed, and there is no point at which we must stop the analysis. Consider, for example, the statement "It is logically necessary that either all men are mortal, or else some men live forever." We can say that this statement has the form "$\Box P$." In other words, it says that something is logically necessary. But we can go further in our analysis of the statement by analyzing the form of the statement symbolized by "P." "P" stands for a disjunction, so we can say that our statement has the form "$\Box(Q \vee R)$." We can go still further, and describe the nature of the disjuncts of the disjunction. The first disjunct has the form "$(x)(Mx \supset Ox)$," and the second disjunct has the form "$(\exists x)(Mx \,\&\, Fx)$." Consequently, we can say that the statement has the form

$$\Box[(x)(Mx \supset Ox) \vee (\exists x)(Mx \,\&\, Fx)]$$

We need not stop here. "F," meaning "lives forever," can be analyzed as "there is a time t such that, for any time t_1 after time t, x is alive at time t_1." Thus "Fx" can be further analyzed as

$$(\exists t)[Tt \,\&\, (t_1)((Tt_1 \,\&\, At_1t) \supset Lxt_1)]$$

Then the original statement has the form

$$\Box\{(x)(Mx \supset Ox) \vee (\exists x)(Mx \;\&\; (\exists t)[Tt \;\&\; (t_1)((Tt_1 \;\&\; At_1t) \supset Lxt_1)])\}$$

We can go still further and analyze "Ox" as

$$(\exists t)(\exists t_1)[Tt \;\&\; Tt_1 \;\&\; At_1t \;\&\; Lxt \;\&\; {\sim}Lxt_1]$$

Then we can say that our original statement has the form

$$\Box\{(x)[Mx \supset (\exists t)(\exists t_1)(Tt \;\&\; Tt_1 \;\&\; At_1t \;\&\; Lxt \;\&\; {\sim}Lxt_1)] \vee (\exists x)(Mx \;\&\; (\exists t)[Tt \;\&\; (t_1)((Tt_1 \;\&\; At_1t) \supset Lxt_1)])\}$$

Nor do we have to stop here. We might go on to analyze "is a man" in terms of something else, and "is a time" in terms of something further, and so forth.

This all suggests that a statement does not have one unique form, but rather can be said to be of many forms. In order to show that a particular argument involving a statement is sound, it may suffice to observe that the statement is a disjunction, while to show that another argument is sound we may have to observe that it is a disjunction of two universal generalizations, and so on. For any given purpose, we may only have to analyze the statement to a certain point, but for another purpose we may have to analyze the statement still further. In discussing the propositional and predicate calculi, we talked about symbolizing the forms of statements, but now we must regard what we were doing there as analyzing statements as far as was possible using the particular concepts symbolized by the logical symbols of those particular theories.

IV.3 EXERCISES

A. Show that in that restricted propositional calculus in which the only sentential connectives are "&" and "$\vee$," there are no valid sentences.

B. Show that in the quantifier free predicate calculus, all valid sentences are tautologies.

C. Should the sentence $(\Box P \supset P)$ be valid in any one of the following.

1. alethic modal logic
2. epistemic logic
3. doxastic logic
4. deontic logic

D. Which of the following sentences are valid in the second order predicate calculus:

1. $(X^0)(\exists Y^0)(X^0 \supset Y^0)$
2. $(\exists Y^0)(X^0)(X^0 \supset Y^0)$
3. $(X^0)(X^0 \vee \sim X^0)$
4. $[(X^1)(X^1a \supset X^1b) \supset (Ga \supset Gb)]$
5. $(\exists X^1)(\exists x)(y)(X^1xy \equiv \sim X^1yy)$
6. $(X_1^1)(Y^1)(\exists Z^1)(x)[Z^1x \equiv (X_1^1x \vee Y^1x)]$

E. The truth rules for universal and existential generalizations in the predicate calculus must be changed somewhat if we are to work with interpretations of the language rather than interpretations of individual sentences. Why? How must they be changed?

IV.4 SYLLOGISTIC

Prior to the twentieth century, logicians generally concerned themselves with syllogistic.[2] Syllogistic can be viewed as part of the predicate calculus. A *categorical proposition* is any statement of the form ⌜All A are B⌝, ⌜Some A are B⌝, ⌜Some A are not B⌝, or ⌜No A are B⌝, where A and B are different predicates, and neither predicate is the negation of the other. As we have seen, these statements can be symbolized in the predicate calculus as $(x)(Ax \supset Bx)$, $(\exists x)(Ax \,\&\, Bx)$, $(\exists x)(Ax \,\&\, \sim Bx)$, and $(x)(Ax \supset \sim Bx)$, respectively. A *syllogism* is any argument (or argument form) having two premises, and a conclusion that is not equivalent to either premise, all of which are categorical propositions.[3] For example, the following are syllogisms:

> All stockbrokers are bald.
> No bald men have dandruff.
> ———
> Therefore, no stockbrokers have dandruff.

> Some peacocks are brilliantly colored.
> All brilliantly colored things are female.
> ———
> Therefore, some peacocks are female.

Syllogistic is concerned with the evaluation of syllogisms.

[2] The theory that we are calling *syllogistic* here is not quite the same as that studied by earlier logicians.

[3] The term *syllogism* is used ambiguously to talk about both the argument and the argument form.

Using what we know about the predicate calculus, we already have tools at hand that allow us to quickly evaluate the soundness of different syllogisms. For example, suppose we want to know whether the following syllogism is sound:

All *A* are *B*.
All *B* are *C*.

Therefore, all *A* are *C*.

Symbolizing this argument form, we get

1. $(x)(Ax \supset Bx)$
2. $(x)(Bx \supset Cx)$

---

3. $(x)(Ax \supset Cx)$

The conclusion of this argument form follows quickly from the premises using the following derivation:

(1)	1.	$(x)(Ax \supset Bx)$	P
(2)	2.	$(x)(Bx \supset Cx)$	P
(1)	3.	$(Aa \supset Ba)$	I, 1 (I16)
(2)	4.	$(Ba \supset Ca)$	I, 2 (I16)
(1, 2)	5.	$(Aa \supset Ca)$	I, 4, 5 (I12)
(1, 2)	6.	$(x)(Ax \supset Cx)$	UG, 5

To take another example, let us consider the first syllogism on the previous page. That can be symbolized as follows:

1. $(x)(Ax \supset Bx)$
2. $(x)(Bx \supset \sim Cx)$

---

3. $(x)(Ax \supset \sim Cx)$

The derivation of this conclusion from its premises is precisely analogous to the above derivation.

To take a further example, let us consider the second syllogism above. We can symbolize it as follows:

1. $(\exists x)(Px \;\&\; Cx)$
2. $(x)(Cx \supset Fx)$

---

3. $(\exists x)(Px \;\&\; Fx)$

Again, the derivation of this conclusion from its premises is straightforward:

(1)	1.	$(\exists x)(Px \,\&\, Cx)$	P
(2)	2.	$(x)(Cx \supset Fx)$	P
(3)	3.	$(Pa \,\&\, Ca)$	P
(2)	4.	$(Ca \supset Fa)$	I, 2 (I16)
(3)	5.	Ca	I, 3 (I2)
(2, 3)	6.	Fa	I, 4, 5 (I9)
(3)	7.	Pa	I, 3 (I1)
(2, 3)	8.	$(Pa \,\&\, Fa)$	I, 6, 7 (I13)
(2, 3)	9.	$(\exists x)(Px \,\&\, Fx)$	I, 8 (I17)
(2)	10.	$[(Pa \,\&\, Ca) \supset (\exists x)(Px \,\&\, Fx)]$	C, 3, 9
(2)	11.	$(x)[(Px \,\&\, Cx) \supset (\exists x)(Px \,\&\, Fx)]$	UG, 10
(2)	12.	$[(\exists x)(Px \,\&\, Cx) \supset (\exists x)(Px \,\&\, Fx)]$	R, 11 (E32)
(1, 2)	13.	$(\exists x)(Px \,\&\, Fx)$	I, 1, 12 (I9)

Although the derivation of the conclusion of a sound syllogism from its premises is generally very simple, it can often be simplified still further by employing a technique that allows transformation of complex syllogisms into simpler syllogisms. To do this, we must first observe that there are certain simple equivalences that hold between categorical propositions:

$$(x)(Ax \supset Bx) \text{ eq. } \sim(\exists x)(Ax \,\&\, \sim Bx)$$
$$(x)(Ax \supset \sim Bx) \text{ eq. } \sim(\exists x)(Ax \,\&\, Bx)$$

We say that $\ulcorner$All A are $B\urcorner$ and $\ulcorner$Some A are not $B\urcorner$ are *contradictories* of one another (they are equivalent to each other's negations) and $\ulcorner$No A are $B\urcorner$ and $\ulcorner$Some A are $B\urcorner$ are contradictories of one another. If φ is a categorical proposition, let us symbolize its contradictory as $\ulcorner\tilde{\varphi}\urcorner$. Clearly $\tilde{\varphi}$ eq. $\sim\varphi$.

Using the above equivalences, together with what is called *the principle of the antilogism,* we can show that certain syllogisms are equivalent to one another. What we mean by saying that two syllogisms are equivalent is that one is sound if, and only if, the other is sound.

The principle of the antilogism is the following tautological equivalence:

$$[(\varphi \,\&\, \theta) \supset \psi] \text{ eq. } [(\varphi \,\&\, \sim\psi) \supset \sim\theta]$$

This is easily demonstrated. $[(\varphi \,\&\, \theta) \supset \psi]$ is equivalent, by exportation, to $[\varphi \supset (\theta \supset \psi)]$, which is equivalent, by contraposition, to $[\varphi \supset (\sim\psi \supset \sim\theta)]$, which is equivalent, by importation, to

$$[(\varphi \,\&\, \sim\psi) \supset \sim\theta].$$

By virtue of this equivalence, if φ, θ, and ψ are three sentences of the predicate calculus, then $[(\varphi \mathbin{\&} \theta) \supset \psi]$ is valid if, and only if, $[(\varphi \mathbin{\&} {\sim}\psi) \supset {\sim}\theta]$ is valid. In particular, if φ, θ, and ψ are categorical propositions, then the corresponding conditional of the syllogism $\ulcorner\varphi$; θ; therefore $\psi\urcorner$ is valid if and only if the corresponding conditional of the syllogism $\ulcorner\varphi$; $\tilde{\psi}$; therefore $\tilde{\theta}\urcorner$ is valid. Consequently, the two syllogisms are equivalent. Let us look at an example of this. Consider the syllogism we examined on page 202.

1. $(\exists x)(Px \mathbin{\&} Cx)$
2. $(x)(Cx \supset Fx)$

---

3. $(\exists x)(Px \mathbin{\&} Fx)$

The contradictory of 1 is $(x)(Px \supset {\sim}Cx)$ and the contradictory of 3 is $(x)(Px \supset {\sim}Fx)$. Hence this syllogism is equivalent to

1. $(x)(Px \supset {\sim}Fx)$
2. $(x)(Cx \supset Fx)$

---

3. $(x)(Px \supset {\sim}Cx)$

This syllogism is obviously sound. 2 is equivalent to $(x)({\sim}Fx \supset {\sim}Cx)$ by contraposition, and 3 follows from this and 1 by hypothetical syllogism (together with appropriate treatment of the quantifiers). Thus making this transformation can considerably shorten the process of showing that the syllogism is sound.

Using this technique to establish the equivalence of syllogisms, every *sound* syllogism can be converted into a syllogism whose premises and conclusion are all universally quantified. Let us see how this goes. We can group all syllogisms not already in this form into four categories:

1. Suppose a syllogism has two universal premises and an existential conclusion. No syllogism of this form can be sound. The reason is that universally quantified categorical propositions are hypothetical in nature. They do not commit us to the existence of anything. $(x)(Ax \supset Bx)$ does not tell us that there are any A's or that there are any B's. It merely says that *if* there are any A's, then they are also B's. On the other hand, an existentially quantified categorical proposition does commit us to the existence of something. Thus it can never follow from two universal premises.

2. Suppose a syllogism has one universal premise, one existential premise, and an existential conclusion. Such a syllogism has the form

$$\begin{array}{l} (x)(Ax \supset Bx) \\ (\exists x)(Cx \,\&\, Dx) \\ \hline (\exists x)(Ex \,\&\, Fx) \end{array}$$

By the above methods, this can be converted into

$$\begin{array}{l} (x)(Ax \supset Bx) \\ (x)(Ex \supset \sim Fx) \\ \hline (x)(Cx \supset \sim Dx) \end{array}$$

3. Suppose a syllogism has two existential premises and a conclusion that is either universal or existential. Such a syllogism cannot be sound because no third categorical proposition can follow from two existential ones. The reason is that two existential categorical propositions cannot be combined to yield anything else, because each says that something exists, but it might be a different thing in each case. For example, the syllogism

$$\begin{array}{l} (\exists x)(Ax \,\&\, Bx) \\ (\exists x)(Bx \,\&\, Cx) \\ \hline (\exists x)(Ax \,\&\, Cx) \end{array}$$

is unsound. The first premise says that there is *something*, let us call it a, which is both A and B. The second premise says there is something, let us call it b, which is both B and C. But a and b need not be the same object, so we cannot conclude that there is anything that is both A and C. For analogous reasons, *no* syllogism having two existential premises is sound.
4. Suppose a syllogism has a universal and an existential premise, and a universal conclusion. Then it has the form

$$\begin{array}{l} (x)(Ax \supset Bx) \\ (\exists x)(Cx \,\&\, Dx) \\ \hline (x)(Ex \supset Fx) \end{array}$$

This syllogism can be converted into

$$\frac{\begin{array}{l}(\exists x)(Ex \ \& \ {\sim}Fx)\\(\exists x)(Cx \ \& \ Dx)\end{array}}{(\exists x)(Ax \ \& \ {\sim}Bx)}$$

Thus this case reduces to case 3, and we see that no syllogism of this form is sound.

Thus all sound syllogisms can be reduced to syllogisms having two universal premises and a universal conclusion. We might summarize our results so far by saying that *if a syllogism is sound, then it has an existential conclusion if, and only if, it has exactly one existential premise.*

Now let us turn to those syllogisms that do have two universal premises and a universal conclusion. Let us define a *standard syllogism* to be any syllogism that has one of the following two forms:

$$\frac{\begin{array}{l}(x)(Ax \supset Bx)\\(x)(Bx \supset Cx)\end{array}}{(x)(Ax \supset Cx)}$$

or

$$\frac{\begin{array}{l}(x)(Bx \supset Cx)\\(x)(Ax \supset Bx)\end{array}}{(x)(Ax \supset Cx)}$$

The only difference between the two forms is that the order of the premises is reversed. The predicate B is called the *middle term* of the standard syllogism. In a standard syllogism it is allowed that any or all of the predicates may be negations of other predicates. For example,

$$\frac{\begin{array}{l}(x)(Ax \supset {\sim}Bx)\\(x)({\sim}Bx \supset {\sim}Cx)\end{array}}{(x)(Ax \supset {\sim}Cx)}$$

is a standard syllogism. Clearly, any standard syllogism is sound. The conclusion follows from the premises by hypothetical syllogism.

It turns out that *given a syllogism that is not a standard syllogism but has two universal premises and a universal conclusion, it is sound if and only if it can be transformed into a standard syllogism by either: (1) contraposing the conclusion; or (2) contraposing one premise; or (3) contraposing the conclusion and one premise.* For example, consider the syllogism

No *B* are *A*.
All non-*B* are *C*.

No non-*C* are *A*.

This is symbolized as

$$\frac{\begin{array}{l}(x)(Bx \supset {\sim}Ax)\\(x)({\sim}Bx \supset Cx)\end{array}}{(x)({\sim}Cx \supset {\sim}Ax)}$$

By contraposing the first premise and the conclusion we get

$$\frac{\begin{array}{l}(x)(Ax \supset {\sim}Bx)\\(x)({\sim}Bx \supset Cx)\end{array}}{(x)(Ax \supset Cx)}$$

which is a standard syllogism. Thus the initial syllogism is sound.

On the other hand consider the syllogism

All *A* are *B*.
All *A* are *C*.

All *B* are *C*.

This is symbolized as

$$\frac{\begin{array}{l}(x)(Ax \supset Bx)\\(x)(Ax \supset Cx)\end{array}}{(x)(Bx \supset Cx)}$$

This syllogism cannot be transformed into a standard syllogism by contraposition. The premises do not have the form necessary for a standard syllogism, because the middle term occurs in the antecedent of both premises. Nor can we rectify this difficulty by contraposing one premise, because then the middle term would be negated in that premise. Thus this syllogism is not sound.

We have not *proven* that every sound syllogism having universal premises and a universal conclusion can be transformed into a standard syllogism by contraposition, but only asserted that they can. In order to prove that this is the case we would have to examine all sixty-four of the different syllogisms that have universal premises and conclusion and show that the sound ones can be thus transformed and that the others are unsound. This is quite easy to do in each case, but rather tedious because of the number of syllogisms involved, so it will not be done here. The skeptical student is invited to do it himself.

This provides a simple way of testing an arbitrary syllogism to determine whether it is sound. It must have either two universal premises and a universal conclusion, or one existential premise and an existential conclusion, to be sound. If it does not we know immediately that it is unsound. If it has an existential premise and existential conclusion, then we can transform it into a syllogism having two universal premises and a universal conclusion by using the principle of the antilogism. Then we can determine whether it is sound by seeing if it can be transformed into a standard syllogism by contraposition. This is often a simpler way of determining whether a syllogism is sound than by trying to give a derivation of the conclusion from the premises or an interpretation under which the premises are true and the conclusion false.

IV.4 EXERCISES

A. Symbolize each of the following syllogisms in the predicate calculus. If they are sound, give derivations of their conclusions from their premises; if they are unsound, find interpretations under which their premises are true and their conclusions false.

1. Some *A* are *B*; some *B* are *C*; therefore, some *A* are *C*.
2. No *A* are *C*; all *B* are *C*; therefore, no *A* are *B*.
3. All *A* are *B*; all *B* are *C*; therefore, some *A* are *C*.
4. No *A* are *B*; no *B* are *C*; therefore, no *A* are *C*.
5. No *A* are *B*; some *A* are *C*; therefore, some *C* are not *B*.
6. No *A* are *B*; all *C* are *A*; therefore, no *C* are *B*.

B. Test the soundness of the syllogisms in A by trying to transform them into standard syllogisms.

IV.5 THE PREDICATE CALCULUS WITH IDENTITY

A Limitations of the Predicate Calculus

The predicate calculus is sufficient to handle most arguments, but there are some rather important statement forms and argument forms that cannot be symbolized in the predicate calculus. For example, we cannot symbolize the form of the statement "Given any positive integer, there is another positive integer greater than it." The difficulty here is with the word "another." Taking the universe of discourse to be the set of positive integers, we might attempt to symbolize this as $(x)(\exists y)Gyx$, but this will not work because we have not said that y is an integer *other than x*. Another statement we cannot symbolize is "John has red hair, but nobody else in the room does." Here we cannot symbolize "nobody else."

Another class of statements whose forms cannot be symbolized in the predicate calculus are those involving *definite descriptions*. A definite description is a phrase of the form "the . . ." which purports to uniquely describe some individual thing. For example, in "The girl with the red hair has had too many martinis," the phrase "The girl with the red hair" is a definite description. We might symbolize such statements by replacing definite descriptions with individual constants, but that would eliminate part of the structure of the statements. For example, "The girl with red hair does not have red hair" is self-contradictory, but if we symbolized "The girl with the red hair" using the individual constant a, this statement would be symbolized as simply $\sim Ra$, which is not inconsistent.

A further very important class of statements whose forms cannot be symbolized in the predicate calculus are numerical statements. For example, we cannot symbolize the statements "There are at least three men in this room who have black hair" or "There are four rings missing from the jewelry box."

We can deal with all of these statement forms if we add a new symbol to the predicate calculus to symbolize identity. The resulting logical theory is called *the predicate calculus with identity*.

B Identity

We say that two objects are identical when they are one and the same object. For example, FDR was the thirty-second President of the United States, so we can say that FDR is identical with the thirty-second President of the United States. In this sense of "identical," to say that two objects are identical is *not* to say merely that they have certain properties in common, but rather to say that they are actually the same object. This sense of "identical" is called "numerical identity," and is to be contrasted with "qualitative identity." For example, if two people have just bought new cars, and the cars are the same make, model, and color, and have precisely the same accessories, we might say that their cars are identical. Here we would mean that they are qualitatively identical—they have their most important qualities or properties in common. We would not mean that they are numerically identical—that would require that the two people had both bought the same car.

Borrowing from mathematics, we will use the equality sign, "=," to symbolize numerical identity. Thus we may write "FDR = the thirty-second President of the United States." The addition of "=" to our formal language as a new logical symbol means that we must extend our concept of an atomic formula. In addition to the atomic formulas of the ordinary predicate calculus, we will take all formulas of the form $\ulcorner(\alpha = \beta)\urcorner$, where α and β are individual symbols (variables or individual constants), to be atomic formulas. Then formulas and sentences of the predicate calculus with identity are built up out of this enlarged list of atomic formulas.

Using identity, we can now symbolize the statements we could not symbolize before. For example, consider the statement "Given any positive integer, there is another positive integer greater than it." This can be paraphrased as "Given any positive integer x, there is a positive integer y which is not identical with x and is greater than x." Taking our universe of discourse to be the set of positive integers, we can then symbolize this as $(x)(\exists y)[\sim(x = y) \,\&\, Gyx]$.

A handy abbreviation, again borrowed from mathematics, is to write $(\alpha \neq \beta)$ for $\sim(\alpha = \beta)$. This will not be a formula of the predicate calculus with identity, but we can use it to abbreviate formulas of the predicate calculus with identity. Using this abbreviation, we can write the above sentence as $(x)(\exists y)[(x \neq y) \,\&\, Gyx]$.

To take another example, let us see how we can symbolize the statement "John has red hair, but nobody else in the room does." This can be paraphrased as "John has red hair, and given anyone in the room x, if x is not John then x does not have red hair." Taking our universe of discourse to be people in the room, this can be symbolized as $\{Rj \,\&\, (x)[(x \neq j) \supset \sim Rx]\}$.

C *Interpretations and Validity*

To make the concepts of validity, implication, and equivalence precise for the predicate calculus with identity, we must again define the notion of an interpretation and give the conditions under which sentences are true under an interpretation. But fortunately, there is very little here that is different from the ordinary predicate calculus. The concept of an interpretation remains unchanged. The only thing we must do regarding truth under an interpretation is explain when a sentence of the form $(\alpha = \beta)$ is true. All other cases are covered by the same truth rules as in the ordinary predicate calculus.

The truth rule for identity is quite simple and really quite obvious. *An identity,* $(\alpha = \beta)$, *is true under an interpretation if, and only if, the interpretation assigns the same denotation to the two individual constants* α *and* β, that is, if and only if the two individual constants denote the same thing. For example, let us consider an interpretation whose domain is $\{1, 2\}$, and which assigns the denotation 1 to a, 2 to b, and 1 to c. Under this interpretation, a and c have the same denotation, so $(a = c)$ is true. But a and b have different denotations, so $(a = b)$ is false.

As in the case of the ordinary predicate calculus, a sentence of the predicate calculus with identity is said to be *valid* if and only if it is true under every interpretation. One sentence *implies* another if, and only if, the corresponding conditional is valid. Two sentences are *equivalent* if, and only if, each implies the other.

D Derivations

We can also construct a system of derivations in the predicate calculus with identity. Using the same rules of inference as in the ordinary predicate calculus, we simply add two new implications to our list of implication schemes. First we add the following implication scheme:

$$\text{(I18)} \quad (\alpha = \beta), Sb\binom{\alpha}{\gamma}\varphi \ \Box\!\!= Sb\binom{\beta}{\gamma}\varphi$$

provided γ is the only variable occurring free in φ. The intuitive idea behind this implication is that if two objects are identical, and one of them has a certain property φ, then so must the other (as it is the same object). For example, from "FDR was the thirty-second President of the United States" and "FDR was bald," we can infer "The thirty-second President of the United States was bald."

The reason I18 holds is that, as we have seen, a formula with free variables symbolizes the form of a predicate or relation. If φ has the single free variable γ, then φ symbolizes the form of a predicate. Predicates are extensional, so if $Sb\binom{\alpha}{\gamma}\varphi$ is true (that is, φ is true of α), and α and β denote the same object (that is, $(\alpha = \beta)$ is true), then $Sb(^{\beta}_{\gamma})\varphi$ must also be true (that is, φ is true of β).

The second implication we must add looks rather strange. Any sentence of the form $(\alpha = \alpha)$ is valid, because under any interpretation, α denotes the same thing as α, making $(\alpha = \alpha)$ true. We want to be able to introduce $(\alpha = \alpha)$ on any line of a derivation, and without premise numbers because it is valid. In order to do this, we may say that $(\alpha = \alpha)$ is implied by the empty set of premises. That is, any interpretation which makes every sentence in the empty set true (which is any interpretation, because there are not any sentences in the empty set) will also make $(\alpha = \alpha)$ true (because it is valid). Thus we can say that the following is an implication:

(I19) $\Box\!\!= (\alpha = \alpha)$

This implication has no antecedent. Using it with Rule I we can introduce a sentence of the form $(\alpha = \alpha)$ on any line of a derivation and have no premise numbers for that line. This may be stretching our original notion of an implication somewhat, but it is simpler to do this than to introduce a new rule for the sole purpose of allowing us to write $(\alpha = \alpha)$ on a line of a derivation.

Now let us prove two simple theorems about identity. First we will show that $(x)(y)[(x = y) \supset (y = x)]$ is valid by giving a derivation of it:

(1)	1.	$(a = b)$	P
	2.	$(a = a)$	I, (I19)
(1)	3.	$(b = a)$	I, 2, 3 (I18) [letting P be $(x = a)$]
	4.	$[(a = b) \supset (b = a)]$	C, 1, 3
	5.	$(y)[(a = y) \supset (y = a)]$	UG, 4
	6.	$(x)(y)[(x = y) \supset (y = x)]$	UG, 5

Next let us show that $(x)(y)(z)\{[(x = y) \& (y = z)] \supset (x = z)\}$ is valid by giving a derivation of it:

(1)	1.	$[(a = b) \& (b = c)]$	P
(1)	2.	$(a = b)$	I, 1 (I1)
(1)	3.	$(b = c)$	I, 1 (I2)
	4.	$(a = a)$	I, (I19)
(1)	5.	$(b = a)$	I, 2, 4 (I18)
(1)	6.	$(a = c)$	I, 5, 3 (I18) [letting *P* be $(x = c)$]

7.	$\{[(a = b) \& (b = c)] \supset (a = c)\}$	C, 1, 6
8.	$(z)\{[(a = b) \& (b = z)] \supset (a = z)\}$	UG, 7
9.	$(y)(z)\{[(a = y) \& (y = z)] \supset (a = z)\}$	UG, 8
10.	$(x)(y)(z)\{[(x = y) \& (y = z)]$ $\supset (x = z)\}$	UG, 9

It is fairly easy to show that derivations in the predicate calculus with identity are complete. To prove this we begin exactly as we did in Section III.16 to prove completeness for the ordinary predicate calculus. We define the set A just as we did there. But we define our interpretation somewhat differently. Rather than assigning a different denotation to each individual constant (as we did there), we only assign a new integer as the denotation of an individual constant α if there was no individual constant β appearing earlier in the list of constants such that $(\alpha = \beta) \in A$. If there was such a constant, then we assign the same denotation to α as we did to β. It is easily verified that, with this modification, we construct an interpretation under which all sentences in A are true. The proof of this is left as an exercise for the student.

E Numerical Statements

An important class of statements that can be dealt with in the predicate calculus with identity are numerical statements. Let us begin by symbolizing statements of the form ⌜There are at least n things that are F⌝. We can handle the case where n is one without using identity. "There is at least one thing that is F" means simply $(\exists x)Fx$. Next consider the statement "There are at least two things that are F." This can be paraphrased as "There is something x, and there is something y, such that x and y are different things and Fx and Fy." This can then be symbolized as

$$(\exists x)(\exists y)[(x \neq y) \& Fx \& Fy]$$

Analogously, "There are at least three things that are F" can be paraphrased as "There is something x, there is something y, and there is something z, such that x, y, and z, are different things, and Fx and Fy and Fz." This can then be symbolized as

$$(\exists x)(\exists y)(\exists z)[(x \neq y) \& (x \neq z) \& (y \neq z) \& Fx \& Fy \& Fz]$$

In general, we can symbolize the statement ⌜There are at least n things that are F⌝ as

$$
\begin{aligned}
(\exists x_1)(\exists x_2)\ldots(\exists x_n)[(x_1 \neq x_2) \;\&\; (x_1 \neq x_3) \;\&\; \ldots \;\&\; (x_1 \neq x_n)\\
\&\; (x_2 \neq x_3) \;\&\; \ldots \;\&\; (x_2 \neq x_n)\\
\vdots\\
\&\; (x_{n-2} \neq x_{n-1}) \;\&\; (x_{n-2} \neq x_n)\\
\&\; (x_{n-1} \neq x_n)\\
\&\; Fx_1 \;\&\; Fx_2 \;\&\; \ldots \;\&\; Fx_n]
\end{aligned}
$$

In other words, we use n existential quantifiers, say that the n variables refer to different things, and say that those things are all F. Let us abbreviate the above sentence as $(\exists_{\geq n}x)Fx$. Thus, $(\exists_{\geq 1}x)Fx$ is $(\exists x)Fx$, and $(\exists_{\geq 2}x)Fx$ is $(\exists x_1)(\exists x_2)[(x_1 \neq x_2) \;\&\; Fx_1 \;\&\; Fx_2]$.

Next consider how we might symbolize statements of the form ⌜There are exactly n things that are F⌝. Let us begin with "There is exactly one thing that is F." This can be paraphrased as "There is something, x, such that Fx, and given anything y, if Fy then y is the same thing as x." This can then be symbolized as

$$(\exists x)[Fx \;\&\; (y)(Fy \supset y = x)]$$

Next consider the statement "There are exactly two things that are F." This can be paraphrased as "There is something x, and there is something y, such that x and y are different things, and Fx and Fy, and given anything z, if Fz then z is either x or y." This can then be symbolized as

$$(\exists x)(\exists y)[(x \neq y) \;\&\; Fx \;\&\; Fy \;\&\; (z)(Fz \supset [(z = x) \vee (z = y)])]$$

In general, in symbolizing the statement ⌜There are exactly n things that are F⌝, we begin just as we did in symbolizing ⌜There are at least n things that are F⌝ and say that there are n things, all different from one another, and all of which are F, but then we add that given anything y, if y is F then y is one of those n things. In other words, we write:

$$
\begin{aligned}
(\exists x_1)(\exists x_2)\ldots(\exists x_n)\{(x_1 \neq x_2) \;\&\; (x_1 \neq x_3) \;\&\; \ldots \;\&\; (x_1 \neq x_n)\\
\&\; (x_2 \neq x_3) \;\&\; \ldots \;\&\; (x_2 \neq x_n)\\
\vdots\\
\&\; (x_{n-2} \neq x_{n-1}) \;\&\; (x_{n-2} \neq x_n)\\
\&\; (x_{n-1} \neq x_n)\\
\&\; Fx_1 \;\&\; Fx_2 \;\&\; \ldots \;\&\; Fx_n\\
\&\; (y)(Fy \supset [(y = x_1) \vee (y = x_2)\\
\vee \ldots \vee (y = x_n)])\}
\end{aligned}
$$

Let us abbreviate this sentence as $(\exists_n x)Fx$.

We can now symbolize the forms of most numerical statements. For example, suppose we want to symbolize the form of the statement "There are three suspicious looking characters in the next room." We can paraphrase this as "There are exactly three things that are suspicious looking characters and are in the next room." This can be symbolized as $(\exists_3 x)(Sx \;\&\; Rx)$.

Next consider the statement "There are more than four things missing." This can be paraphrased as "There are at least five things missing," which can be symbolized as $(E_{\geq 5}x)Mx$.

Finally, consider the statement "There are less than three men in the world who know how to perform this operation." This can be paraphrased as "It is not the case that there are at least three men in the world who know how to perform this operation," and then symbolized as $\sim(\exists_{\geq 3}x)(Mx \;\&\; Wx \;\&\; Kx)$.

F Definite Descriptions

Statements involving definite descriptions are more difficult to symbolize. Although there is no uniform agreement on just how they are to be symbolized, the most common way to symbolize them was devised by Bertrand Russell, and that will be the course followed here.

Consider the statement "The girl with the red hair has had too many martinis." Following Russell, we can paraphrase this statement as "There is exactly one girl with red hair (in our universe of discourse), and she has had too many martinis." In other words, when we say "the girl with the red hair," this commits us to there being someone who is picked out by this definite description, that is, to there being exactly one girl with red hair in our universe of discourse, and then we go on to say something about the person thus identified. We can further paraphrase this statement as "There is someone *x* such that *x* is a girl with red hair (thus saying that there is at least one girl with red hair) and *x* has had too many martinis, and given anyone *y*, if *y* is a girl with red hair then *y* is *x* (thus saying there is only one girl with red hair)." This can be symbolized as

$$(\exists x)\{Gx \;\&\; Hx \;\&\; Mx \;\&\; (y)[(Gx \;\&\; Hx) \supset (y = x)]\}$$

Another way we might paraphrase this is "There is exactly one girl with red hair, and every girl with red hair has had too many martinis." We can use the universal quantifier in the second conjunct because we have already said that there is exactly one such girl. This statement can then be symbolized as $\{(\exists_1 x)(Gx \;\&\; Hx) \;\&\; (y)[(Gy \;\&\; Hy) \supset My]\}$. We might also paraphrase this statement as "There is exactly one girl with red hair, and there is a girl with red hair who has had too many martinis." We can use the existential quantifier in the second conjunct because we have already said that there is exactly one such girl. Thus the statement can be symbolized as

$$[(\exists_1 x)(Gx \ \& \ Hx) \ \& \ (\exists y)(Gy \ \& \ Hy \ \& \ My)]$$

All three symbolizations of this statement are equivalent, so it makes no difference which symbolization we use.

It is customary to introduce a new symbol for symbolizing statements involving definite descriptions. We can symbolize the definite description ⌜the thing α that is F⌝ as ⌜$(\iota\alpha F\alpha)$⌝. This is called the *description operator*. Then ⌜The thing α that is F has the property G⌝ can be symbolized as ⌜$G(\iota\alpha F\alpha)$⌝. We will introduce this as an abbreviation rather than as a new symbol. $G(\iota\alpha F\alpha)$ will be short for

$$\ulcorner(\exists\alpha)\{F\alpha \ \& \ G\alpha \ \& \ (\beta)[F\beta \supset (\beta = \alpha)]\}\urcorner$$

Then, for example, the symbolization of "The girl with the red hair has had too many martinis" can be abbreviated as $M(\iota x(Gx \ \& \ Hx))$.

Notice one peculiarity about the expression ⌜$(\iota\alpha F\alpha)$⌝. We use it to simplify the writing of long formulas, but ⌜$(\iota\alpha F\alpha)$⌝ is not itself an abbreviation for something else. Rather, the whole formula ⌜$G(\iota\alpha F\alpha)$⌝ is an abbreviation for something else. In explaining the meaning of ⌜$(\iota\alpha F\alpha)$⌝, we give a rule for eliminating it from any formula in which it occurs rather than giving a definition of it itself in terms of some longer formula. In other words, we give a definition of ⌜$G(\iota\alpha F\alpha)$⌝ rather than just of ⌜$(\iota\alpha F\alpha)$⌝.

One class of statements involving definite descriptions is the class of identity statements. For example, we may say "Carol is the girl with the red hair." Using our symbol for definite descriptions, the statement will look like this: $[c = (\iota x(Gx \ \& \ Hx))]$. Following the above method of symbolizing, and letting Gx be $(c = x)$, we can symbolize this statement as $(\exists x)\{(Gx \ \& \ Hx) \ \& \ (c = x) \ \& \ (y)[(Gy \ \& \ Hy) \supset (y = x)]\}$. But clearly, as we say that $c = x$, we can replace x throughout this sentence with c, giving us $\{(Gc \ \& \ Hc) \ \& \ (y)[(Gy \ \& \ Hy) \supset (y = c)]\}$. This is the simplest way to symbolize an identity statement one side of which is a definite description. In general, a statement of the form ⌜α is the β such that $F\beta$⌝ will be symbolized as

$$\ulcorner\{F\alpha \ \& \ (\beta)[F\beta \supset (\beta = \alpha)]\}\urcorner$$

It can be abbreviated as ⌜$[\alpha = (\iota\beta F\beta)]$⌝.

Notice that in a particular universe of discourse there may not be any such thing as "the thing that is F." For example, if our universe of discourse is the universal set, then there is no such thing as either "the girl with red hair" or "the unicorn with purple spots," in the former case because there is more than one girl with red hair and in the latter case because there are no unicorns with purple spots. To say that there is such a thing as "the thing that is F" is just to say that there is something x such that x is the thing that is F, that is, $(\exists x)[x = (\iota x Fx)]$. This in turn is an abbreviation for

$$(\exists x)[Fx \;\&\; (y)(Fy \supset (y = x))]$$

which is just $(\exists_1 x)Fx$. Hence, to say that there is such a thing as "the thing that F" is just to say that there is exactly one thing that is F.

There is one class of statements involving definite descriptions that creates additional difficulties. These statements are those in which the definite descriptions involve *superlatives*, such as "The highest mountain," or "the greatest president," or "the oldest man in the room." In general, these definite descriptions have the form ⌜the R-est thing that is F⌝ (for example, "the oldest thing that is a man in this room"). In general, ⌜the R-est thing that is F⌝ can be paraphrased as ⌜the thing α such that α is F and given anything else β, if β is F then α is R-er (for example, "older") than β⌝. For example, "the oldest man in this room" can be paraphrased as "the thing x such that x is a man in this room and given anything else y, if y is a man in this room then x is older than y." The definite description ⌜the R-est thing that is F⌝ can then be symbolized as

$$\ulcorner(\iota\alpha(F\alpha \;\&\; (\beta)\{[(\beta \neq \alpha) \;\&\; F\beta] \supset R\alpha\beta\}))\urcorner$$

Applying this to a specific case, let us see how we would symbolize the form of the statement "The highest mountain in Marin County is 1400 feet tall." Letting M stand for "is a mountain in Marin County," H for "(1) is higher than (2)," and T for "is 1400 feet tall," we can first paraphrase this statement as

T(the highest mountain in Marin County)

and then as

T(the highest thing that is a mountain in Marin County)

and then as

$$T(\iota x(Mx \;\&\; (y)\{[(y \neq x) \;\&\; My] \supset Hxy\}))$$

This can be spelled out as the abbreviation for

$$(\exists x)[Mx \,\&\, (y)\{[(y \neq x) \,\&\, My] \supset Hxy\} \,\&\, Tx \\ \&\, (z)([Mz \,\&\, (y)\{[(y \neq z) \,\&\, My] \supset Hzy\}] \supset (z = x))]$$

Unfortunately, a rather perplexing difficulty arises in connection with this treatment of definite descriptions. This concerns the specification of the predicate G in a formula of the form $\ulcorner G(\iota\alpha F\alpha)\urcorner$. The difficulty can be illustrated by considering the difference between the statements "The king of France is not bald" and "It is not the case that the king of France is bald." The former statement is analogous to the statement "The king of France is bald." That is, it can be analyzed as "There is exactly one person who is the king of France, and he is not bald." This is symbolized as $[(\exists_1 x)Kx \,\&\, (\exists x)(Kx \,\&\, \sim Bx)]$. On the other hand, the statement "It is not the case that the king of France is bald" is the negation of the statement "The king of France is bald," and so is to be symbolized as $\sim[(\exists_1 x)Kx \,\&\, (\exists x)(Kx \,\&\, Bx)]$, which is equivalent to $[(\exists_1 x)Kx \supset (x)(Kx \supset \sim Bx)]$. There is an important difference between these two sentences. The first says that there is a king of France and that he is not bald. The second merely says that *if* there is a king of France then he is not bald. The first implies that there is a king of France while the second does not. The difficulty that arises is that we do not yet have any way to show the difference between these two statements by symbolizing them using the description operator. It would be natural to symbolize both of them as $\sim B(\iota xKx)$. But this is an abbreviation for

$$[(\exists_1 x)Kx \,\&\, (\exists x)(Kx \,\&\, \sim Bx)]$$

Thus $\sim B(\iota xKx)$ is only a correct symbolization for the first of the above two statements. Another way to point out the difficulty is by observing that $\sim B(\iota xKx)$ is not the negation of $B(\iota xKx)$. Rather, $\sim B(\iota xKx)$ is the abbreviation for the formula we get by replacing G by $\sim B$ in $G(\iota xKx)$ and expanding it according to our rule for eliminating definite descriptions. Perhaps the clearest way to say this is that in $\sim B(\iota xKx)$, if we follow our rule for eliminating definite descriptions, it is the predicate B that is negated rather than the whole sentence. It is for this reason that $\sim B(\iota xKx)$ is not a proper symbolization of "It is not the case that the king of France is bald."

We want to have a way of symbolizing statements containing definite descriptions both when the whole statement is negated and when only the predicate is negated. To do this we must have a way of indicating how much of the sentence that results from symbolizing the statement is to be counted as part of the predicate, so that we know whether the negation is to apply only to the predicate or to the whole sentence. The method adopted here will be to enclose the predicate in square brackets and then follow it by the definite description. Thus we will write $[Gx](\iota xFx)$. In the above case, if we want to negate only the predicate we will write $[\sim Bx](\iota xKx)$, and if we want to negate the whole sentence we will write $\sim[Bx](\iota xKx)$. This can be generalized to predicates of arbitrary complexity. For example, if K stands for "is king of France," B stands for "has black hair," and D stands for "dyes his hair," then $[Bx \supset Dx](\iota xKx)$ means "There is exactly one person who is king of France, and if he has black hair then he dyes it." On the other hand, $([Bx](\iota xKx) \supset [Dx](\iota xKx))$ means "If there is exactly one person who is king of France and he has black hair, then there is exactly one person who is king of France and he dyes his hair," which is equivalent to "If there is exactly one person who is king of France, then if he has black hair he dyes it." Thus the former sentence implies that there is a king of France, but the latter does not.

We must now reexamine our rule for eliminating definite descriptions from sentences. The rule only applies to the case of eliminating them from expressions of the form $[Gx](\iota xFx)$ where the formula in square brackets is atomic. Stating our rule in a more general form, we can say that *if* φ *and* θ *are formulas of the predicate calculus that contain free occurrences of some variable* α *and* θ *does not contain free occurrences of any other variables, then* $\ulcorner[\varphi](\iota\alpha\theta)\urcorner$ *is an abbreviation for the formula*

$$\ulcorner[(\exists_1\alpha)\theta \;\&\; (\exists\alpha)(\varphi \;\&\; \theta)]\urcorner$$

However, this is still not the most general form of the rule. This form can only be used in formulas that contain a single definite description. But some formulas contain more than one definite description. For example, we might symbolize "The girl with the red hair is better looking than the girl with the black hair" as

$$[Lxy]((\iota xRx), (\iota yBy))$$

Our rule for eliminating definite descriptions does not tell us which of these descriptions to eliminate first, but the order of elimination will make a difference to what sentence we obtain after the elimination. If we eliminate (ιxRx) first we will obtain the sentence

$$\{(\exists_1 x)Rx \;\&\; (\exists x)[Rx \;\&\; (\exists_1 y)By \;\&\; (\exists y)(By \;\&\; Lxy)]\}$$

whereas if we eliminate $(\imath yBy)$ first we will obtain

$$\{(\exists_1 y)By \ \& \ (\exists y)[By \ \& \ (\exists_1 x)Rx \ \& \ (\exists x)(Rx \ \& \ Lxy)]\}$$

However, $[Lxy]((\imath xRx), (\imath yBy))$ can only be an abbreviation for one of these two sentences. We must formulate our elimination rule in such a way that it determines the order of elimination of definite descriptions from formulas containing more than one definite description. Fortunately, it does not make any difference how we do this, because the different sentences we obtain using different orders of elimination are all equivalent to one another. Consequently, we will adopt the simplest rule, which is to eliminate the definite descriptions in the order in which they are listed: *if φ and $\theta_1, \ldots, \theta_n$ are formulas of the predicate calculus, and each of $\theta_1, \ldots, \theta_n$ contains a free occurrence of the variable $\alpha_1, \ldots, \alpha_n$ respectively but not of any other variables, and φ contains free occurrences of all of $\alpha_1, \ldots, \alpha_n$, then $\ulcorner[\varphi]((\imath\alpha_1\theta_1), \ldots, (\imath\alpha_n\theta_n)\urcorner$ is an abbreviation for the formula*

$$\ulcorner[(\exists_1\alpha_1)\theta_1 \ \& \ldots \& \ (\exists_1\alpha_n)\theta_n \ \& \ (\exists\alpha_1)\ldots(\exists\alpha_n)(\varphi \ \& \ \theta_1 \ \& \ldots \& \ \theta_n)]\urcorner$$

For example, $[L\imath xy]((xRx), (\imath yBy))$ becomes an abbreviation for

$$[(\exists_1 x)Rx \ \& \ (\exists_1 y)By \ \& \ (\exists x)(\exists y)(Lxy \ \& \ Rx \ \& \ By)]$$

which is equivalent to both of the above tentative symbolizations of it.

When we write $\ulcorner[\varphi](\imath\alpha\theta)\urcorner$, we say that φ is the *scope* of the definite description. English sentences are frequently ambiguous in just this respect. It is often unclear what the scope of a definite description is supposed to be. For example, consider the English sentence "Charles de Gaulle is not the king of France." This could be interpreted either as $\sim[g = x](\imath xKx)$ or $[g \neq x](\imath xKx)$. The former means "It is not the case that there is a king of France who is identical with Charles de Gaulle," that is, "If there is a king of France then he is not identical with Charles de Gaulle," while the latter means "There is a king of France, but he is not identical with Charles de Gaulle." The difference in scope is just the difference between whether or not the sentence implies that there is a king of France. In any given case, we must rely upon contextual clues to determine the scope of the definite description. For example, a person who said today that Charles de Gaulle is not the king of France would most likely not mean to imply that there is a king of France, and so his statement could be symbolized as $\sim[g = x](\imath xKx)$. But this is not something we can know just by looking at the sentence itself. Of course, we have run into similar problems in the ordinary predicate calculus. Most English sentences are to some extent ambiguous, but the ambiguity is generally easier to resolve than in the case of the scope of definite descriptions.

Sometimes the scope notation can be omitted without ambiguity. This will be the case whenever the scope of the definite description is an atomic formula. For example, we can write $\sim G(\imath xFx)$ rather than $\sim [Gx](\imath xFx)$. Whenever the scope is not explicitly indicated, this means that it is the smallest possible scope, that is, it consists of the atomic formula in which the definite description occurs. Thus if we write $\sim G(\imath xFx)$ this can never be mistaken for $[\sim Gx](\imath xFx)$ because the latter can never be written without using the scope notation. This simplification allows us to use definite descriptions just like individual constants in constructing formulas. For example, we can write

$$\{Fa(\imath xGx) \supset \sim [Fab \equiv (b = (\imath xGx))]\}$$

Here the scope of the first occurrence of the definite description is $[Fax]$ and the scope of the second occurrence is $[b = x]$.

IV.5 EXERCISES

A. Symbolize the following statements in the predicate calculus with identity. For those that involve definite descriptions, first symbolize them using the description operator, and then eliminate the description operator to obtain sentences of the predicate calculus with identity. In each case, give a justification for your choice of the scope of the definite description.

1. No two people are exactly the same height.
2. There are at least three ways to symbolize the statement on line one.
3. Wild Bill Hickock is the fastest draw west of the Mississippi.
4. The President of the United States is not senile.
5. Mr. Merriweather has two sons, and one of them is taller than the other.
6. Herbert Feenelhoffer is not the first man to swim the Atlantic Ocean.
7. Albert Einstein was not the first man to swim the English Channel.
8. No one but the President can prevent war.
9. The least accurate clock is more accurate than the most accurate sundial.

10. Sarah's pie was not the best pie in the contest.
11. Two is the only even prime number.
12. The girl with the red hair despises the girl with the black hair, and is trying to steal her boyfriend.

    (Symbolize the following using F for "(1) is a father of (2)," M for "(1) is a mother of (2)," a for "Arthur," and p for "Priscilla.")

13. Priscilla is the paternal grandmother of Arthur.
14. Arthur is the father of Priscilla's maternal grandmother.

B. Show that "The girl with the red hair does not have red hair" is inconsistent.

C. Show that $[(\imath xFx) = (\imath xGx)], H(\imath xFx) \models H(\imath xGx)$.

D. Establish that the sentences in the following pairs are equivalent to one another by deriving each sentence in each pair from the other member of the pair:

1. $(\exists_2 x)Fx$; $[(\exists_{\geq 2} x)Fx \;\&\; \sim(\exists_{\geq 3} x)Fx]$
2. $(\exists x)[Fx \;\&\; Gx \;\&\; (y)(Fy \supset (y = x))]$; $[(\exists_1 x)Fx \;\&\; (y)(Fy \supset Gy)]$
3. $[(\exists_1 x)Fx \;\&\; (y)(Fy \supset Gy)]$; $[(\exists_1 x)Fx \;\&\; (\exists y)(Fy \;\&\; Gy)]$
4. $(\exists x)[Fx \;\&\; (x = c) \;\&\; (y)(Fy \supset (y = x))]$; $[Fc \;\&\; (y)(Fy \supset (y = c))]$

IV.6 FORMAL AXIOMATIC THEORIES

Most students are familiar with the axiomatization of Euclidean plane geometry; it is an example of a formal axiomatic theory. In a formal axiomatic theory, we take certain sentences of a formal language as *axioms,* and then the theory consists of the axioms together with all sentences that are implied by the axioms. The sentences that are implied by the axioms are called the *theorems* of the theory. A formal axiomatic theory is always based on an underlying logical theory which supplies the relation of implication. Those formal theories that are based on the predicate calculus or the predicate calculus with identity are called *first-order theories.* For example, if we began by symbolizing all of the statements of plane geometry in the predicate calculus and then adopted Euclid's axioms as the axioms of the formal axiomatic theory, we would have a first-order theory whose theorems were just the truths of Euclidean plane geometry. First-order theories are of particular interest because derivations in the predicate calculus give us a mechanical way of deriving all of the theorems from the axioms (that is, the predicate calculus is effectively generable). This is not true, for example, of theories based on the second-order predicate calculus, because there is no complete system of derivations for that logical theory.

There are many reasons why we may want to axiomatize a set of statements. An axiomatization of a set of beliefs or a body of knowledge will tell us just how much or how little must be assumed before everything else follows, because in order to get an adequate axiomatization we must make all of our assumptions explicit. And if there is some question about whether some particular statement is true, the matter can be settled if we can show that it follows from axioms already known to be true. Furthermore, as we shall see, an axiomatization of a body of knowledge or set of beliefs will often make it much easier to show that the basic assumptions of the body of knowledge or set of beliefs are consistent with one another.

Let us consider an example of a formal axiomatic theory. Consider the set of all true statements about the nonnegative integers (the integers zero, one, two, and so on) in which the only relations employed are "less than" and "less than or equal to." If we symbolize these statements in the predicate calculus with identity, using L for "(1) is less than (2)" and M for "(1) is less than or equal to (2)," we can axiomatize this set of statements using the following axioms:

A1. $(x)(y)(Lxy \supset \sim Lyx)$

A2. $(x)(y)(Lxy \vee Lyx \vee x = y)$

A3. $(x)(y)(z)[(Lxy \,\&\, Lyz) \supset Lxz]$

A4. $(x)(\exists y)Lxy$

A5. $(\exists x)(y)Mxy$

A6. $(x)(y)[Mxy \equiv (Lxy \vee x = y)]$

A1 says that given any two nonnegative integers, if the first is less than the second, then the second is not less than the first. A2 says that given any two nonnegative integers, either they are equal or one is less than the other. A3 says that given any three nonnegative integers, if the first is less than the second and the second is less than the third, then the first is less than the third. A4 says that given any nonnegative integer, there is a larger nonnegative integer. A5 says that there is a smallest nonnegative integer (namely, zero). A6 is just the definition of the "less than or equal to" relation.

From these axioms we can derive a number of theorems, and a few examples follow. In order to shorten the derivations, they are abbreviated somewhat. Several steps may be carried out on a single line, and although rules of inference are cited for each line, no mention is made of which implication or equivalence schemes are used. By this time the student should be sufficiently familiar with the predicate calculus to be able to follow these abbreviated derivations.

Theorem 1: $(x) \sim Lxx$

(1)	1.	$(x)(y)(Lxy \supset \sim Lyx)$	A1
(1)	2.	$(Laa \supset \sim Laa)$	I, 1
(1)	3.	$\sim Laa$	I, 2
(1)	4.	$(x) \sim Lxx$	UG, 3

Theorem 2: $(x)(y)(z)[(Mxy \,\&\, Myz) \supset Mxz]$

(1)	1.	$(Mab \,\&\, Mbc)$	P
(2)	2.	$(x)(y)[Mxy \equiv (Lxy \vee x = y)]$	A6
(2)	3.	$[Mab \equiv (Lab \vee a = b)]$	I, 2
(1, 2)	4.	$(Lab \vee a = b)$	I, 1, 3
(2)	5.	$[Mbc \equiv (Lbc \vee b = c)]$	I, 2
(1, 2)	6.	$(Lbc \vee b = c)$	I, 1, 5
(1, 2)	7.	$[(Lab \vee a = b) \,\&\, (Lbc \vee b = c)]$	I, 4, 6
(1, 2)	8.	$\{[(Lab \vee a = b) \,\&\, Lbc]$ $\vee [(Lab \vee a = b) \,\&\, b = c]\}$	R, 7
(9)	9.	$[(Lab \vee a = b) \,\&\, Lbc]$	P
(9)	10.	$[(Lab \,\&\, Lbc) \vee (a = b \,\&\, Lbc)]$	R, 9
(11)	11.	$(Lab \,\&\, Lbc)$	P
(12)	12.	$(x)(y)(z)[(Lxy \,\&\, Lyz) \supset Lxz]$	A3
(12)	13.	$[(Lab \,\&\, Lbc) \supset Lac]$	I, 12
(11, 12)	14.	Lac	I, 11, 13
(16)	15.	$(a = b \,\&\, Lbc)$	P
(16)	16.	Lac	I, 15
(9, 12)	17.	Lac	I, 10, 14, 16
(9, 12)	18.	$(Lac \vee a = c)$	I, 17
(19)	19.	$[(Lab \vee a = b) \,\&\, b = c]$	P
(19)	20.	$(Lac \vee a = c)$	I, 19
(1, 2, 12)	21.	$(Lac \vee a = c)$	I, 8, 18, 20
(2)	22.	$[Mac \equiv (Lac \vee a = c)]$	I, 2
(1, 2, 12)	23.	Mac	I, 21, 22
(2, 12)	24.	$[(Mab \,\&\, Mbc) \supset Mac]$	C, 1, 23
(2, 12)	25.	$(x)(y)(z)[(Mxy \,\&\, Myz) \supset Mxz]$	UG, 24

Theorem 3: $(x)Mxx$

Theorem 4: $(x)(y)(z) \sim (Lxy \,\&\, Lyz \,\&\, Lzx)$

Theorem 5: $(x)(y)[(Mxy \,\&\, Myx) \supset x = y]$

Theorem 6: $(x)(y)(Lxy \supset x \neq y)$

Theorem 7: $(x)(y)(Mxy \vee Myx)$

Theorem 8: $(x)(\exists y)(Mxy \,\&\, x \neq y)$

Theorem 9: $(\exists x)(y)Mxy$

Theorem 10: $(x)(y)[Lxy \equiv (Mxy \,\&\, x \neq y)]$

What particular axioms we choose to axiomatize a certain set of statements will often be a matter simply of convenience. There is nothing intrinsic about a formula that makes it an axiom. We make something an axiom by *choosing* to take it as such, and if we wished we could choose other formulas as axioms and still get the same set of theorems. For example, instead of taking A1–A6 as axioms, we could have taken Theorems 2, 5, 7–10 as axioms, and still have obtained exactly the same theorems. In order to show that this is the case, it suffices to show that these two sets of axioms are derivable from one another. Then any sentence derivable from one of the sets of axioms is derivable from the other. Theorems 2, 5, 7–10 are derivable from axioms A1–A6, so all that remains is to show that axioms A1–A6 are derivable from Theorems 2, 5, 7–10. This is left as an exercise for the student.

This example of a formal axiomatic theory indicates the need for a certain qualification to our definition of the concept of a theorem of a theory. In constructing a theory, we generally only assign meanings to a few relation symbols and individual constants. In the example above we assigned meanings to L and M, but to no other relation symbols and no individual constants. Our axioms are formulated using only these relation symbols, so they have a determinate meaning. However, there will be sentences implied by the axioms that use other relation symbols too, and these sentences will not have a determinate meaning. For example, $[(x)(\exists y)Lxy \vee (\exists x)Fx]$ is implied by A4, but as we have assigned no meaning to F we are not interested in this sentence. In general, we are only interested in those sentences which contain no relation symbols or individual constants not contained in the axioms. So let us redefine "theorem" accordingly: a sentence is a *theorem* of a formal axiomatic theory if, and only if, it is implied by the axioms and contains no nonlogical symbols not contained in the axioms. Let us say that the *language of the theory* consists of all sentences that contain no nonlogical symbols not contained in the axioms. Then, the theorems are all sentences of the language of the theory that are implied by the axioms.

Three concepts of fundamental importance in discussing formal axiomatic theories are those of consistency, soundness, and completeness. A formal axiomatic theory is *consistent* if, and only if, it is not the case that some of the axioms imply the negation of another axiom. For a first-order theory this is the same thing as requiring that it is possible for all of the axioms to be true simultaneously, that is, that there is an interpretation (in the predicate calculus or predicate calculus with identity) under which all of the axioms are true. Notice that, as the theorems are implied by the axioms, if an interpretation makes all of the axioms true, it will also make all of the theorems true. A formal axiomatic theory is *inconsistent* if, and only if, it is not consistent. Clearly, in order for it to be of any use as an axiomatization of a body of knowledge or set of truths, a theory must be consistent. If it is inconsistent, then the theorems cannot all be true simultaneously, and so the theorems do not symbolize a set of true statements.

The concepts of a sound theory and a complete theory are always relative to a particular assignment of meanings to the individual constants and relation symbols (or in theories that are not first-order theories, an assignment of meanings to the nonlogical symbols). A formal axiomatic theory is generally intended to axiomatize a certain preselected set of true statements, and thus we will have a particular assignment of meanings in mind when we are dealing with the theory —namely, that assignment of meanings under which the sentences of the language of the theory symbolize the statements in which we are interested. For example, for the formal axiomatic theory considered above, the interpretation we had in mind was to let L stand for "(1) is less than (2)" and M stand for "(1) is less than or equal to (2)." This assignment of meanings that we have in mind when we construct a theory is called the *standard assignment of meanings for the theory*. For a first-order theory the standard assignment of meanings generates what we call the *standard interpretation of the theory*—that interpretation which assigns the denotations, extensions, and so forth, dictated by the standard assignment of meanings. When we are axiomatizing a set of true statements, the set of theorems that we want our theory to have are just all those sentences that, under the standard assignment of meanings, symbolize the *true* statements among the statements that we are symbolizing. For a first-order theory, these are just those sentences that are true under the standard interpretation. This leads us to define a *sound* theory to be a theory under which all of the theorems are true under the standard assignment of meanings, and a *complete* theory to be a theory under which all of the sentences true under the standard assignment of meanings are theorems. For first-order theories we can state this somewhat more precisely by saying that the theory is sound if, and only if, all of the theorems are true under the standard interpretation, and complete if, and only if, all of the sentences true under the standard interpretation are theorems. Thus a sound theory is one that does not contain any theorems that we do not want, and a complete theory is one that does contain all the theorems that we do want. Clearly, in order to be a proper axiomatization of a body of truths, a theory must be both sound and complete.

These concepts of soundness and completeness are related to, but not the same as, the concepts of soundness and completeness used in discussing logical theories. In each case, a sound theory is one that does not contain any theorems that we do not want, and a complete theory is one that does contain all theorems that we do want. But in the case of logical theories, it is valid sentences that we want as theorems, while in the case of formal axiomatic theories it is just true sentences (under the standard interpretation) that we want as theorems.

In discussing formal axiomatic theories, it is rather natural to ask to what extent we can axiomatize the various branches of mathematics. Let us look at ordinary arithmetic. This consists of statements involving "+," "·," and the numerals "0," "1," "2," and so on, and quantifiers ranging over the nonnegative integers. Our underlying logical theory will be the predicate calculus with identity. We will let a_0 stand for "0," and a_1 for "1." The other integers can be defined in terms of these two integers. For example, "3" can be defined as "(1 + 1) + 1." Consequently, it is not necessary to have individual constants denoting them. P will stand for the relation "(1) plus (2) is equal to (3)," and T will stand for "(1) multiplied by (2) is equal to (3)." Our universe of discourse will be the set of nonnegative integers.

In axiomatizing the true statements of arithmetic, we first need axioms that tell us that addition and multiplication are functions, that is, that the addition of two numbers or the multiplication of two numbers always gives just one value:

A1. $(x)(y)(\exists_1 z)P(x, y, z)$

A2. $(x)(y)(\exists_1 z)T(x, y, z)$

In order to make the language of our theory look more familiar, let us write $x + y = z$ as an abbreviation for $P(x, y, z)$, and $x \cdot y = z$ as an abbreviation for $T(x, y, z)$. We can also write "0" as an abbreviation for a_0, "1" for a_1, "2" for $\imath x(1 + 1 = x)$, "3" for $\imath x(2 + 1 = x)$, "4" for $\imath x(3 + 1 = x)$, and so on. We will also write formulas of the form $\ulcorner F(\alpha + \beta)\urcorner$ as an abbreviation for $\ulcorner F(\imath\delta(\alpha + \beta = \delta))\urcorner$. For example, the sentence $(x)(y)(z)[x + (y + z) = (x + y) + z]$ is an abbreviation for

$$(x)(y)(z)[x + \imath x_1(y + z = x_1) = \imath x_2(x + y = x_2) + z]$$

which in turn is an abbreviation for

$$(x)(y)(z)(\exists_1 x_1)(\exists_1 x_2)[P(y, z, x_1) \;\&\; P(x, y, x_2) \;\&\; P(x, x_1, (x_2 + z))]$$

This finally is an abbreviation for

$$(x)(y)(z)(\exists_1 x_1)(\exists_1 x_2)(\exists_1 x_3)[P(y, z, x_1) \;\&\; P(x, y, x_2) \\ \;\&\; P(x_2, z, x_3) \;\&\; P(x, x_1, x_3)]$$

Next we need axioms characterizing the number sequence. From A1 we can derive the conclusion $(x)(\exists y)x + 1 = y$, that is, for every integer there is another integer that is one greater. Thus, starting from zero we get a sequence of numbers $0, 0 + 1, (0 + 1) + 1, \ldots$. We want to say that this is a linear sequence, that is, that it does not go around in a circle. For this purpose we take the two axioms

A3. $\sim(\exists x)0 = x + 1$

A4. $(x)(y)(x + 1 = y + 1 \supset x = y)$

The first of these axioms tells us that we can never get back to zero by adding one. The second axiom tells us that we can never get back to any other number that the sequence has already passed by adding one, that is, if $x + 1$ is some number, and by adding one to y we get the same number, then y must be the same number as x, and hence we cannot get back to $x + 1$ by adding one to some larger number.

Finally, in order to characterize the number sequence, we need an axiom that tells us that the numbers we get by starting from zero and successively adding one are all the numbers that there are. We can approximate this goal by using the axiom scheme of induction:

A5. $\{[F(0) \,\&\, (x)(F(x) \supset F(x + 1))] \supset (x)F(x)\}$

This is an axiom scheme rather than a single axiom. That is, any sentence that we can get by substituting some formula of the language of this theory for F is an axiom. This axiom scheme tells us that if some predicate is true of zero, and true of every number that can be obtained by successively adding one to zero, then it is true of all numbers. Clearly, this axiom will be true if these are all the numbers there are. In turn, if F could be *any* predicate, this axiom scheme would imply that these *are* all the numbers there are, for let F be the predicate, "can be obtained from zero by successively adding one." Then the antecedent of the conditional is true, and so the consequent must be true. But the consequent means "Every number can be obtained from zero by successively adding one." However, it is not clear how to define the predicate "can be obtained from the zero by successively adding one" within the language of this theory. Thus, although the axiom scheme of induction is clearly true, it is not clear whether it is sufficient for saying that all numbers can be obtained from zero by successively adding one.

Finally, we need axioms characterizing addition and multiplication:

A6. $(x)x + 0 = x$

A7. $(x)(y)x + (y + 1) = (x + y) + 1$

A8. $(x)x \cdot 0 = 0$

A9. $(x)(y)x \cdot (y + 1) = (x \cdot y) + x$

These axioms are called the *recursive definitions* of addition and multiplication. They allow us to calculate the value of any number added to any other number, or any number multiplied by any other number. For example, let us look at A6 and A7. For each number x, A6 tells us what the result of adding the smallest number, 0, to x is. Then for each larger number $y + 1$ (a number larger than zero is always something plus one), A7 tells us what x plus that number is in terms of the value of x plus the next smallest number. For example, we can calculate the value of $3 + 4$ as follows:

$3 + 4$	$= 3 + (((1 + 1) + 1) + 1)$	(by definition of "4")
	$= (3 + ((1 + 1) + 1)) + 1$	(by A7)
	$= ((3 + (1 + 1)) + 1) + 1$	(by A7)
	$= (((3 + 1) + 1) + 1) + 1$	(by A7)
	$= ((4 + 1) + 1) + 1$	(by definition of "4")
	$= (5 + 1) + 1$	(by definition of "5")
	$= 6 + 1$	(by definition of "6")
	$= 7$	(by definition of "7")

Clearly these axioms are all true under the standard interpretation, so the formal axiomatic theory is sound. The more difficult question is whether it is complete. There was once strong reason to think that it was complete, because it was known that if we used the second order predicate calculus rather than the ordinary predicate calculus as our underlying logical theory, and replaced axiom scheme A5 by the single axiom

A5. $(X^1)\{[X^1(0) \,\&\, (x)(X^1(x) \supset X^1(x + 1))] \supset (x)X^1(x)\}$

the resulting theory was complete. However, in 1931 Kurt Gödel proved, to the amazement of most logicians, that the above first-order theory was incomplete. This naturally suggests that we add some additional axioms in order to make it complete. However, Gödel proved the even stronger result that this theory cannot be made complete by adding any finite set of additional axioms or axiom schemes.[4] This is surprising, because it means that the arithmetic of the natural numbers cannot be axiomatized in a way that will allow us to mechanically derive all true statements from the axioms. The axiomatization within the second order predicate calculus does not accomplish this, because, as was noted before, it is impossible to construct a complete system of derivations for the second order predicate calculus. This latter result, incidentally, is also a consequence of Gödel's remarkable theorem.

IV.6 EXERCISES

1. Show that the formal axiomatic theory discussed on pages 222–224 is consistent.
2. Show that if a theory is sound, then it is consistent.

[4] Gödel's theorem is actually a bit stronger than this, but it is impossible to state it precisely without a great deal more preparation.

3. Show that if a first-order theory is inconsistent, then every sentence of the language of the theory is a theorem.
4. Show that if a first-order theory is inconsistent, then it is complete.

IV.7 THE NATURE OF ANALYTICITY

In explaining what logic is all about, we have relied heavily upon the concept of an analytic statement, but that concept itself is not very clear. It is easy enough to give examples of analytic statements and contrasting examples of nonanalytic statements, and in this way the concept can be explained to a person not previously familiar with it. And some things can be said about analytic statements to help clarify their nature. We can point out that analytic statements do not just happen to be true—they *must* be true—they are *necessarily* true. We cannot even imagine what it would be like for an analytic statement to be false. For example, it is impossible to conceive of there being a married bachelor, or a brother who is not a sibling. Analytic statements seem to lay down the limits of the concepts in terms of which we think of the world. The analytic statement that all bachelors are unmarried tells us something about the concept of a bachelor, and in virtue of that we cannot even think of someone as being a bachelor without thinking of him as being unmarried. We might say that analytic truths lay down the conceptual framework that we use to describe the world.

Still left unanswered, however, is the question of how statements get to be analytic, or precisely what it means to say that a statement is analytic. Philosophers down through the ages have suggested a number of answers to this question, and perhaps it will clarify things a bit to examine the predominant views about analyticity.

a *The Contingency Theory.* According to this theory, analytic statements are just very general truths about the world, differing from scientific laws and other truths only in that they are more general. We discover an analytic statement just like we discover any other general truth about the world—by observing the world and seeing that the statement always holds. For example, we discover the (nonanalytic) law of gravity by noticing that whenever the support is removed from something, it falls to the ground. Analogously, according to this theory, we discover the (analytic) law of the excluded middle (the law that given any statement, it is either true or false) by noticing that whenever we investigate a particular statement, it always turns out to be either true or false.

However, this theory overlooks one of the most important features of analytic statements—we know that they are true without examining the world around us. Our knowledge of the law of gravity is based upon the observation of a number of particular falling bodies, but our knowledge of the law of the excluded middle is not grounded upon any observations of the world. We can somehow simply *see*, independently of any experience of the world, that the law of the excluded middle holds; indeed, we cannot even conceive of what it would be like for it to be false. But we can imagine perfectly well what it would be like for the law of gravity not to hold. Thus it is extremely misleading to compare analytic statements with scientific laws. They are different in a very important respect.

b Psychologism. Analytic statements are statements whose falsity is inconceivable. We cannot even imagine what it would be like for an analytic statement to be false. This suggests that analyticity has something to do with human powers of imagination, and analytic statements accordingly express some sort of "laws of thought." This is the psychologistic theory of analyticity. It gains credence from the fact that the source of our knowledge of analyticity is somehow within us rather than in the world around us. To discover that a statement is analytic we do not have to examine the world, but only reflect upon the meaning of the statement.

The principal difficulty with the psychologistic theory is that analytic statements are true. Their being analytic *makes* them true—in fact, necessarily true. But the psychologistic theory is unable to explain why our inability to imagine the falsity of a statement should make it true. What have human powers of imagination got to do with the way the world is? Furthermore, there is no reason to think that everybody will have the same powers of imagination, or the same abilities to conceive possibilities. Presumably young children and morons cannot imagine as much as other people. Does this mean that what is analytic for one person may not be analytic for another? If so, then, according to the psychologistic theory, there really could not be any connection between analyticity and truth. But it is part of the very concept of an analytic statement that it is necessarily true; so the psychologistic theory cannot constitute a correct account of analyticity.

c Conventionalism. The clearest examples of analytic statements are those that appear to arise directly from definitions. For example, "All brothers are siblings" seems to be analytic by virtue of the definition of "sibling" as meaning "brother or sister." It is tempting to suppose that all analytic statements arise in this way from definitions of words. Language is *conventional.* By this it is meant that there is nothing intrinsic about words that forces them to have just the meanings they do in fact have. The meanings of words are established through an historical process in which people slowly come to use the words in a certain way. But if a significant portion of the linguistic community got together and decided to use a word in another way, there would be nothing to prevent this. In this sense, the meanings of words are determined by *linguistic conventions* that can be changed more or less at will. The conventionalist theory of analyticity suggests that analytic statements merely express these linguistic conventions. "All brothers are siblings" is analytic by virtue of the linguistic convention that leads us to use "sibling" to mean "brother or sister." Analyticity reflects nothing but the way we use words, and does not give us any profound insights into the makeup of the universe. We *make* sentences of our language analytic by choosing to use words in a certain way, and if we wished we could make these same sentences nonanalytic by choosing to use words in a different way.

In order to see what is wrong with conventionalism, we must distinguish between sentences and statements. Sentences are always *of* a particular language. "Snow is white" is a sentence of English, while "Schnee ist weiss" is a sentence of German. However, people speaking different languages can say the same things, or make the same statements. A German saying "Schnee ist weiss" would be making the same statement as an American saying "Snow is white." What is true or false is not the sentence itself, but rather the statement that is made by using the sentence. A contemporary speaker of the English language who said "All brothers are siblings" would be making a certain statement, and this statement is analytic. The conventionalist is quite right that we could if we wished change the meaning of the word "sibling" so that the sentence "All brothers are siblings" would no longer be used to make an analytic statement. If we let "sibling" have the same meaning as "tomato," then "All brothers are siblings" would mean "All brothers are tomatoes," and a person saying this would be making a statement that is nonanalytic, and in fact false. But this would not change the analyticity of the statement that we now make when we say "All brothers are siblings." If at the same time we changed the meaning of "sibling," we changed the meaning of "radish" to mean "brother or sister," then we could express the statement that we now express by saying "All brothers are siblings" by saying "All brothers are radishes," and this statement would still be analytic. Linguistic conventions do not seem to have anything to do with what statements are analytic. They only have to do with the question of which sentences of a particular language can be used to make analytic statements. Thus conventionalism fails to explain what it is that makes a statement analytic.

d Intuitionism. According to the logical intuitionist, we have some sort of unexplained mental faculty that allows us to simply *see* analytic truths. This logical intuition is compared with sight, and it is maintained that just as sight allows us to see relations between objects in the material world, so our logical intuition allows us to see relations between concepts in some sort of logical realm.

The intuitionist theory emphasizes an important aspect of analytic truths—the fact that we can somehow see them directly, without investigating the world around us. We come to know analytic truths simply by reflection. But this in itself is not sufficient to explain what makes statements analytic. To say that we have this unexplained power of logical intuition is simply to record the existence of the phenomenon without attempting to explain it, but what we are seeking is an explanation of analyticity. It appears that the intuitionist is no closer to explaining analyticity than are the proponents of the other theories we have considered.

The contingency theory, psychologism, conventionalism, and intuitionism are the four most important theories to be proposed in an attempt to give an account of analyticity. But, as we have seen, none of them is entirely successful. The exact nature of analyticity is as problematic now as it ever was. This is indeed unfortunate, because the concept of analyticity is one of the corner stones upon which our treatment of logic has been based. However, this is a rather common state of affairs in philosophy. Whenever philosophers have undertaken investigations of the foundations of anything, they have almost always encountered extremely difficult problems for which they have not been able to arrive at any commonly agreed upon solutions.

APPENDIX IMPLICATIONS AND EQUIVALENCES IN THE PREDICATE CALCULUS

A. Truth-functional implications and equivalences:

I1.	$(P \mathbin{\&} Q) \;\square\!\!- P$	(simplification)
I2.	$(P \mathbin{\&} Q) \;\square\!\!- Q$	
I3.	$P \;\square\!\!- (P \vee Q)$	(addition)
I4.	$P \;\square\!\!- (Q \vee P)$	
I5.	$\sim P \;\square\!\!- (P \supset Q)$	
I6.	$Q \;\square\!\!- (P \supset Q)$	
I7.	$\sim(P \supset Q) \;\square\!\!- P$	
I8.	$\sim(P \supset Q) \;\square\!\!- \sim Q$	
I9.	$P, (P \supset Q) \;\square\!\!- Q$	(*modus ponens*)
I10.	$\sim Q, (P \supset Q) \;\square\!\!- \sim P$	(*modus tollens*)
I11.	$\sim P, (P \vee Q) \;\square\!\!- Q$	(disjunctive syllogism)
I12.	$(P \supset Q), (Q \supset R) \;\square\!\!- (P \supset R)$	(hypothetical syllogism)
I13.	$P, Q \;\square\!\!- (P \mathbin{\&} Q)$	(adjunction)
I14.	$(P \vee Q), (P \supset R), (Q \supset R) \;\square\!\!- R$	(dilemma)
I15.	$P_1, \ldots, P_n \;\square\!\!- (P_1 \mathbin{\&} \ldots \mathbin{\&} P_n)$	
E1.	$\sim\sim P$ eq. P	(double negation)
E2.	$\sim(P \vee Q)$ eq. $(\sim P \mathbin{\&} \sim Q)$	(De Morgan's laws)
E3.	$\sim(P \mathbin{\&} Q)$ eq. $(\sim P \vee \sim Q)$	
E4.	$(P \supset Q)$ eq. $(\sim P \vee Q)$	
E5.	$\sim(P \supset Q)$ eq. $(P \mathbin{\&} \sim Q)$	
E6.	$\sim(P \equiv Q)$ eq. $(P \equiv \sim Q)$	
E7.	$(P \equiv Q)$ eq. $[(P \supset Q) \mathbin{\&} (Q \supset P)]$	
E8.	$(P \equiv Q)$ eq. $[(P \mathbin{\&} Q) \vee (\sim P \mathbin{\&} \sim Q)]$	

E9.	$(P \,\&\, (Q \,\&\, R))$ eq. $((P \,\&\, Q) \,\&\, R)$	(commutative laws)
E10.	$(P \vee (Q \vee R))$ eq. $((P \vee Q) \vee R)$	
E11.	$(P \,\&\, Q)$ eq. $(Q \,\&\, P)$	(associative laws)
E12.	$(P \vee Q)$ eq. $(Q \vee P)$	
E13.	$[P \,\&\, (Q \vee R)]$ eq. $[(P \,\&\, Q) \vee (P \,\&\, R)]$	(distributive laws)
E14.	$[P \vee (Q \,\&\, R)]$ eq. $[(P \vee Q) \,\&\, (P \vee R)]$	
E15.	$[P \supset (Q \supset R)]$ eq. $[(P \,\&\, Q) \supset R]$	(exportation—importation)
E16.	$(P \supset Q)$ eq. $(\sim Q \supset \sim P)$	(contraposition)
E17.	$[R \vee (P \,\&\, \sim P)]$ eq. R	
E18.	$(P \vee P)$ eq. P	(idempotency laws)
E19.	$(P \,\&\, P)$ eq. P	
E20.	$[R \,\&\, (P \vee \sim P)]$ eq. R	
E21.	$[R \,\&\, (P \,\&\, \sim P)]$ eq. $(P \,\&\, \sim P)$	

B. Non-truth-functional implications and equivalences:

I16. $(x)P \;\Box\!\!= Sb\binom{a}{x}P$

I17. $Sb\binom{a}{x}P \;\Box\!\!= (\exists x)P$

E22.

$(x)P$ eq. $\sim(\exists x) \sim P$

$(\exists x)P$ eq. $\sim(x) \sim P$

$\sim(x)P$ eq. $(\exists x) \sim P$

$\sim(\exists x)P$ eq. $(x) \sim P$

E23.	$(x)(P \vee Q)$ eq. $[(x)P \vee Q]$	Provided there are no free occurrences of x in Q.
E24.	$(Q \supset (x)P)$ eq. $(x)(Q \supset P)$	
E25.	$((\exists x)P \supset Q)$ eq. $(x)(P \supset Q)$	
E26.	$(x)P$ eq. $(y)Sb\binom{y}{x}P$	Provided y does not occur in P.
E27.	$(\exists x)P$ eq. $(\exists y)Sb\binom{y}{x}P$	

SELECTED BIBLIOGRAPHY

Students interested in pursuing further some of the above topics may find the following works useful. References are suggested for specific chapters and sections, with complete bibliographic data being given at the end of the list.

II. THE PROPOSITIONAL CALCULUS

(5), chapter 1
(2), chapters 2, 3, 7, 8
(7), chapter 1
(1), chapters 2, 3

III. THE PREDICATE CALCULUS

(5), chapters 2, 3
(2), chapters 4, 5, 9
(8), chapters 1–8
(6), chapters 2, 6
(1), chapters 4, 5

IV. OTHER TOPICS IN LOGIC

Section IV.2. Use and mention
(8), chapter 2, section 1

Section IV.3. Logical theories
(3), parts 7, 8, 9
(5), chapters 2, 4
(1), chapter 5

Section IV.4. Syllogistic
(11), part one

Section IV.5. The predicate calculus with identity
(8), chapters 9, 10
(6), chapter 3
(10), chapters 7, 8

Section IV.6. Formal axiomatic theories
- (8), chapter 11
- (9), chapters 3, 4
- (6), chapters 4, 5
- (7), chapters 8–11

Section IV.7. The nature of analyticity
- (4), part eight

(1) Church, Alonzo, *Introduction to Mathematical Logic*, vol. I, Princeton, N. J., 1956.

(2) Copi, Irving, *Symbolic Logic*, 2nd ed., New York, 1965.

(3) Copi, Irving and James Gould, *Contemporary Readings in Logical Theory*, New York, 1967.

(4) Edwards, Paul and Arthur Pap, *A Modern Introduction to Philosophy*, New York, 1957.

(5) Hilbert, D. and W. Ackermann, *Mathematical Logic*, 2nd ed., New York, 1950.

(6) Kleene, Stephen, *Mathematical Logic*, New York, 1967.

(7) Kleene, Stephen, *Introduction to Metamathematics*, Princeton, N. J., 1950.

(8) Mates, Benson, *Elementary Logic*, New York, 1965.

(9) Mendelson, Elliott, *Introduction to Mathematical Logic*, Princeton, N. J., 1964.

(10) Rosser, J. Barkley, *Logic for Mathematicians*, New York, 1953.

(11) Terrell, D. B., *Logic, a Modern Introduction to Deductive Reasoning*, New York, 1967.

INDEX